ALL THE BOATS ON THE OCEAN

ALL THE BOATS ON THE OCEAN

How Government Subsidies Led
to Global Overfishing

CARMEL FINLEY

THE UNIVERSITY OF CHICAGO PRESS
CHICAGO AND LONDON

The University of Chicago Press, Chicago 60637
The University of Chicago Press, Ltd., London

Printed in the United States of America

26 25 24 23 22 21 20 19 18 17   1 2 3 4 5

ISBN-13: 978-0-226-44337-9 (cloth)
ISBN-13: 978-0-226-44340-9 (e-book)
DOI: 10.7208/chicago/9780226443409.001.0001

Library of Congress Cataloging-in-Publication Data

Names: Finley, Carmel, author.
Title: All the boats on the ocean : how government subsidies led to global overfishing / Carmel Finley.
Description: Chicago : The University of Chicago Press, 2017. | Includes bibliographical references and index.
Identifiers: LCCN 2016033849 | ISBN 9780226443379 (cloth : alk. paper) | ISBN 9780226443409 (e-book)
Subjects: LCSH: Fishery management—United States—History—20th century. | Fisheries—North Pacific Ocean—History—20th century. | Fishery management—Political aspects—North Pacific Ocean. | Fishery policy—United States. | Overfishing—North Pacific Ocean. | Sea-power—Economic aspects.
Classification: LCC SH221 .F55 2017 | DDC 333.95/6—dc23 LC record available at https://lccn.loc.gov/2016033849

♾ This paper meets the requirements of ANSI/NISO Z39.48-1992 (Permanence of Paper).

CONTENTS

ACKNOWLEDGMENTS

I started the first draft of this book in an airy office at the Rachel Carson Center in Munich, looking out over a sea of green chestnut trees. The center gave me six months of writing time, and it was the most creative and inspiring setting I could ever hope to have. At the risk of leaving out many, many names, I thank Helmuth Trischler and Christof Mauch, as well as Melanie Arndt, Fiona Cameron, Amy Hay, Andrea Kiss, Ed Russell, Fei Sheng, and Gordon Winder.

I am grateful for the support of Angela Andrea, Mary Elizabeth Braun, Debbra Bacon, Bernice and Bruce Barnett, Holly Campbell, Dianne Cassidy, Lorenzo Ciannelli, Ron Doel, Karin Ellison, Selina Heppell, Ingo Heidbrink, Bob Hitz, Jennifer Hubbard, Mike Jager, Jake Hamblin, Kris Harper, Mary Hunsicker, Linda McPhee, Ellen Pikitch, Hans and Karin Radtke, Bill Robbins, Helen Rozwadowski, Kay Sagmiller, James Sagmiller, Vera Schwach, Martin and Lyudmila Schuster, Jeanie and Tom Senior, Bill Robbins, and Kevin Walsh.

It was San Diego tuna fisherman Captain Dick Stevenson who first told me about the "away boats" that caught tuna in the Pacific for delivery to the canneries in Puerto Rico, then fished in the Atlantic before retracing their steps to return home to California. Deborah Day, the former archivist at Scripps Institute of Oceanography, gave me access to the American Tuna Association (ATA) files. August Felando, the ATA's last director, also shared his insights. The Bez family in Seattle, especially Renee Bez, generously shared family files. Victor Lundquist at the National Oceanic and Atmospheric Administration helped with background and photographs. The Truman Presidential Library found the picture of Harry Truman fishing for salmon and Nick Bez rowing the boat. Thanks also to the University of Washington Special Collections for their access to the Chapman papers.

I am grateful for comments on the manuscript from Sidney Holt, Daniel Pauly, Kevin Bailey, Ellen Pikitch, and David Sampson; and for the editorial assistance of Jeanie Senior and Roberta Ulrich. I appreciate the support of my family, Peter and Nancy MacDougall, and John, Katy, and Solon MacDougall, and my editorial team at Chicago, headed by Christie Henry. I also thank my guys: my husband Carl who read countless drafts, and Max, who purred me through the home stretch. The mistakes are my own.

INTRODUCTION

Political Roles for Fish Populations

A man went looking for his wife, whom the killer whales had abducted. He lifted up the edge of the sea as though it were a blanket, and walked under. As he journeyed on, he came across some very pale looking fish, and, wishing to please them, he painted them all red. And the fish have been red ever since.[1]
—Tlingit legend

Fishing has always been about much more than just catching fish. Fishing is one of the imperial strategies that nation states employed as they struggled for ocean supremacy. Being a seafaring empire required a range of enterprises and fishing was often just a step toward securing other, more desirable objectives. Hugo Grotius was trying to expand the Dutch empire when he wrote *The Free Sea* in 1609. The concept of the Freedom of the Seas has been useful to empires ever since. It made world trade profitable. And in the twentieth century, it was central to the rapid postwar expansion of fishing from a coastal, inshore activity to a global enterprise—one that has been so technologically successful there is literally no place left in the oceans for fish to hide.

Nations rapidly industrialized after World War II, part of the so-called Great Acceleration as people sharply increased their extraction of resources from the natural world.[2] Heady with the scientific and technical knowledge developed during the war, nations dammed rivers, moved mountains, and tried to change ocean currents and alter weather patterns. All of these events created unprecedented changes in the biosphere itself.[3] The ocean has especially been altered. Human activity has changed the ocean, not only close to shore, but as historian Naomi Oreskes argues, "in its *entirety*."[4]

Many fisheries are not sustainable, either biologically or economically. For most of the last decade, it was generally considered that the marine harvest peaked at 86.4 million metric tons (mt) in 1996. A new catch reconstruction suggests the catch could have been as high as 130 million metric tons, and that catches are declining at a steeper rate than scientists originally thought, with just 74.4 million metric tons caught in 2010.[5] The global fish catch costs $105 billion (U.S. dollars) to catch, but it sold for $80 billion.[6] The fishing power of fleets worldwide may be as much as 250 percent higher than what would be needed to fish at ecologically sustainable levels,[7] yet governments continue to subsidize the building of new fishing boats.

Any review of the problems with fishing—and there have been many—agree that too many boats were built, far in excess of the number needed to harvest fish stocks. But the reviewers never ask the question, *who built all the boats in the first place?* The iconic image of the lone fisherman in his dory is so strongly pervasive that the blame for overfishing is placed on fishermen, not on the deliberate government actions in creating policies to greatly expand postwar fisheries. Government money eased the transition between salted, dried, and canned fish, to a world of frozen fish and new products such as fish sticks. Government money hastened the spread of wartime technologies such as radar and sonar, greatly increasing the ability of boats to find and target fish. As colonial empires broke up on land after the war, they were re-created in the oceans, creating a new stage of imperialism. The United States, Japan, and the Soviet Union, as well as the British, Germans, and Spanish, industrialized their fisheries. Nations like South Korea and Communist China, as well as the Eastern bloc countries of Poland and Bulgaria, also began fishing on an almost unimaginable scale. There was a rush for the new and old fishing nations to find new stocks of fish to exploit.

Fishing was a fast way to modernize and industrialize an economy. Shipbuilding, especially on an industrial scale, provided good jobs, boosted coastal economies, and provided fish for export. If your fishermen didn't catch the fish off your coast, fishermen from some other nation would. By the early 1960s, massive fleets of fishing boats were landing more and more fish, and protests were growing about the decline of fish stocks in home waters. Fishing was a territorial claim on the last frontier on the planet: the seas. Fishing was also a form of exerting control over the ocean space as a part of the construction of power in the state system.[8] The race for the oceans, and the fish and whales in them, was a primary battlefield during the Cold War, which lasted from early 1946 until the fall of the Soviet Union in 1990.

Cold War trade dynamics had an enormous impact on the expansion of global fisheries. The expansion of fishing has to be understood in terms of the breadth of the *American* conception of national security after 1948. As political scientist Melvyn Leffler argues, "the conception included a strategic sphere of influence within the western hemisphere, domination of the Atlantic and Pacific oceans, an extensive network of bases to enlarge the strategic frontier and project American power, an even more extensive system of transit rights to facilitate the conversion of commercial air bases to military use, access to the resources and markets of most of Eurasia, denial of those resources to a prospective enemy, and the maintenance of nuclear superiority."[9] Each and every component on Leffler's list had an impact on fish and fishing.

If the problem of world hunger could be solved, it would remove a major cause of social instability. "Americans believed that economic instability and poverty bred political chaos, revolutionary behavior, totalitarianism, violence, aggression, and war," wrote historian Thomas Paterson. "It was assumed that these conditions were attractive to political extremists like Communists who always preyed on weaknesses and dislocations."[10] It was surely possible to increase the harvest from the sea. Fish that were not caught died anyway, of no benefit to mankind.

Many American policy makers believed that developing world trade would bring world peace. If the problem of world hunger could be solved, that would ease the social unrest that made communism attractive in third-world countries. Japan and Germany had to be reintegrated into the global community. Japan, in particular, had to be rebuilt to bolster political stability in Asia and serve as an example for the Russians and Chinese of how successful a capitalist economy could be.[11] The Americans also wanted to strengthen their position in the Pacific by developing an American-style economy in its new possessions: the Mariana, Caroline, and Marshall Islands, soon to be important sites for nuclear testing. The U.S. also sought to anchor its European allies—especially Iceland and Norway—to its interests, militarily and economically. The vehicle to accomplish all these goals was fish, specifically the families of Pacific tuna (*Scombridae*) and Atlantic cod (*Gadus morhua*).

The Cold War was fought on the seas. The development of naval and submarine warfare for both the U.S. and U.S.S.R. greatly accelerated the science of oceanography.[12] Nations also went fishing on a vast scale, developing new technologies to find and process fish. For some countries, like Japan, fishing was one of the foundation stones of the economy. They had been the world's leading fishing nation throughout the 1930s, but the fleet that had always been far too large for its home waters to support. Under the American occupation, the fleet was rapidly rebuilt, in record time. Once a peace treaty was signed in 1951, Japanese fisherman began to set their miles of long lines throughout the North Pacific, hauling in salmon (*Oncorhynchus*), walleyed pollock (*Theragra chalcogramma*), and king crab (*Paralithodes camtschaticus*). Their tuna fleets set millions of hooks in the waters of the southern Pacific, as well as the Indian and Atlantic oceans, seeking bluefin tuna (*Thunnus thynnus*), one of the largest and most valuable fish in the sea, with machine-like efficiency and intensity.

The rapid rebuilding of the Japanese fishing fleet under the American occupation stimulated the industrialization of fisheries in other Asian countries, especially South Korea and Communist China. The Soviet Union, which had long been frustrated by the Japanese salmon catch in Kamchatka and Sakhalin, now restored to Soviet control, was eager to take over the fishing infrastructure Japan had left behind.

The Soviets went fishing on a massive scale. By the mid-1960s, they had fleets of sophisticated boats operating throughout the Atlantic and they were steadily escalating fishing in the Pacific. Fleets from the Black Sea were sailing through the Suez Canal to tap the resources of the Indian Ocean, operating "with the precision of a naval task force."[13] With Soviet assistance, Communist China built a fishing fleet; the catch more than doubled in just four years, from 2.64 million metric tons to 5.6 million metric tons by 1960. The Koreans bought boats from Japanese fishermen, who were using government subsidies to build new boats. The first South Korean exploratory trawler went to the eastern Bering Sea in 1966. Commercial operations began the next year and production reached an estimated 5,000 metric tons of pollock in 1970 and 1971.[14]

Starting in 1961, Spain created the world's third-largest fishing fleet. Within a decade, frozen fish production went from 4,000 metric tons to approximately 500,000 metric tons. Spanish fishermen operated substantial fisheries off the Americas and along the African coast from Tangiers to Capetown.[15] The government provided generous government loans and other subsidies. Cuba wanted to turn Havana into a major Atlantic fishing port. By 1962, Cuba was buying factory trawlers from Spanish shipyards, hiring Soviet captains, and investing heavily to enter the cod fishery.[16]

European and Asian boats were fishing on the West African shelf area, while boats from South Korea and Taiwan not only expanded their coastal fisheries, but also joined the world tuna longline fishery.[17] Much of the fishing was driven by market demand for fish meal, the prime ingredient in the food fed to the modern poultry and livestock industries. With the collapse of California sardines in the 1950s, processing equipment and capital was moved to Peru, creating the largest single-species fishery in the world,

Peruvian anchovies (*Engraulis ringens*).[18] Landings increased 27 percent a year, culminating in a peak catch of 12 million metric tons by 1970 before the fishery collapsed.[19] The equipment was moved to South Africa.

Canada was also embarking on an extensive program of fisheries development, catching fish to export. Newfoundland voted to join the Canadian confederation in 1949, and the federal government began a rich flow of money to modernize fisheries in Newfoundland and the three other Maritime Provinces. The American market was ready to absorb as much cod as fishermen could catch. The government would help improve refrigerated railcars, be involved in a cooperative advertising campaign with the industry, and participate in an extensive program of research into technologies designed to catch more fish.[20] A small-boat insurance program was started in 1953, and the Fisheries Improvement Loans Act was established in 1955 to provide loans to purchase or repair fishing boats and equipment.[21] The Canadian government made fishermen eligible for a special fishermen's unemployment insurance program in 1957, designed to pay benefits to seasonal workers.[22] This program encouraged workers to move into the fishing industry, and to collect payments during the closed seasons.

The American government was building boats, but not for domestic fishermen. Instead, there was a boat-building program to build fishing boats for Cold War allies—specifically the Soviet Union. It took until 1970 before there were low-interest loans and other subsidies to help build an American fishing fleet. In the meantime, as the Cold War deepened, the State Department used concessions around fishing to further two of its political goals: the rebuilding of the Japanese economy, and a closer trade relationship with Iceland, where the U.S. had strategically important military bases.

Fishing was one of the first domestic industries hurt by postwar American trade policy and its emphasis on open markets. For decades, domestic industries like fishing had depended on import tariffs to offset cheap wages in foreign countries and higher American high labor costs. After 1949, as Japanese tuna poured into American supermarkets and Canadian and Icelandic fillets were disrupting the New England industry, fishermen sought relief through tariffs. But their concerns were trumped by the State Department and its unwillingness to upset relations with Japan and Iceland.

While the Americans squabbled over tariff relief through the 1960s, fleets of enormous ships, hundreds of feet long, spread throughout the world's oceans, catching as many fish as they could. Fishing was being revolutionized, and the traditional structure of an industry that was centuries old was being restructured along explicitly political lines, as Soviet factory ships appeared in American waters. The Soviet fishing fleet—and the science it depended on—was one of the great Soviet achievements. Government support meant that instead of a shipyard just building one boat, it would build dozens. The latest technologies spread rapidly, especially to third-world countries, where industrialized nations were eager to do business and sell fishing gear and negotiate the right to catch fish.

As the foreign fishing increased, protests escalated; nations sought to protect their local stocks from the factory processing fleets. New underwater discoveries showed the ocean floor was littered with potentially valuable minerals, and it was surely only a matter of time before scientists would learn how to mine them economically. In the scramble to protect the existing fish resources and the potential mineral deposits, governments in the 1960s started to expand their territorial seas, putting limits on foreign fishing and whaling. They also subsidized the domestic expansion of fishing, often on a massive scale.

The tension between utilizing and protecting the seas led to the Law of the Sea process and the three meetings, in 1957, 1973, and 1982, that created modern maritime law. There is voluminous literature about each of those meetings and the events that led up to them. This account focuses on how the expansion of fisheries in the Pacific during the Cold War stimulated the globalization of fishing and the creation of international fisheries management.

We traditionally tell two stories about the harvest from the seas: one story about whales, the other about fish. Whaling and fishing are considered separate issues (although whaling is technically fishing). They are managed separately, but there is much commonality and intermingling of the science. Fishing and whaling are inextricably linked and ought to be considered together, both by ecologists trying to merge two sets of biological data, and by citizens, who struggle to understand how to better conserve both fish and whales. It was problems faced by nations like Peru (plagued by illegal whaling in its waters) and Iceland (which had fought with foreign fishing boats in its waters since the 1880s) that led nations to create wider territorial seas during the 1970s, as a way to constrain unlimited harvest by foreign fleets in their home waters. The result was the rapid overcapitalization of the global fishing fleet and the subsequent destruction of many stocks.

As this historical reconstruction shows, both Japan and the Soviet Union greatly expanded whaling, as well as fishing, in the North Pacific Ocean and the Bering Sea after World War II. Both expansions were driven by the same economic, political, and social forces. The fish and whales caught were of course important in and of themselves, but fishing and whaling were also territorial claims, as well as a challenge to American naval superiority.

Some scientists believe that industrial whaling in the North Pacific created a legacy of sequential collapse: with the removal of the great whales, the smaller but more deadly killer whales turned to preying on sea otters and sea lions, leading to the collapse of some populations.[23] This analysis is controversial, especially among scientists.[24] But it should be no question that fishing and whaling have their roots in the same Cold War imperatives. The modern factory processing ship, capable of staying at sea for weeks at a time and catching fish by the ton, derived from the whaling industry. These industrial fishing boats played a role in removing up to 90 percent of the largest fish in the sea—another analysis that is controversial, especially among scientists.[25]

The traditional story of the development of fisheries over the centuries is that they radiated outward from Northern Europe and the North Sea, moving west to the New World and from the cold waters of the northern hemisphere to the tropical waters of the south. There is a second story of fishery development, which is only starting to be told, that is the movement of fishing from the Pacific into the Atlantic and the Indian Oceans. This wave of development was much faster, propelled by imperialism and industrialization, the ideology of modernity, and massive amounts of government money.

This book has its roots in my days as a newspaper reporter, working as a correspondent for the *Oregonian* during the 1980s. I was based in Newport, on the Oregon coast, and I wrote a lot of stories about the economic development that was going to come with full utilization of coastal groundfish stocks. One of the stories I vividly remember was a meeting that Oregon Sea Grant agent Bob Jacobson set up with four of Newport's older fishermen. One of them was Gordon White, and he had me enthralled—talking about trawling in the 1940s, heading out of Newport onboard the new trawler *Yaquina*, with just a depth finder to help them find fish. There weren't coastal cities then, there were coastal towns, small towns, and on the ocean in the dark, there weren't many lights to guide them back home.[26] It was Gordon White who first told me about rosefish, or Rosies, and how the local stocks had been destroyed by the Soviet factory processing ships in the late 1960s. Six decades after the Soviet fishery decimated rockfish stocks, they have yet to recover. How did the Soviet boats come to be fishing off Oregon? Why have rockfish stocks not recovered? And how was the depletion of these local fish stocks the result of what had to be national and international events? This book is my answer to those questions.

The scientific name for the bright red fish is *Sebastes alutus*, but there are many other names. Within fishery management they are known as Pacific ocean perch, or POP, but they are marketed as ocean perch or even red snapper. They were once the dominant fish population along the West Coast of the United States, from Mexico to Alaska, although their cousins are found in many other places. They were decimated by the Soviet fishing fleet, first off Alaska, then British Columbia, and finally Oregon and Washington. Within the space of 15 years, rosefish were gone.

The Soviet ships—as well as Bulgarian, West German, Japanese, Chinese, and Polish—all fished in the waters off Canada and the U.S., both on the east and west coasts, until new legislation was enacted in the 1970s creating exclusive economic zones, or EEZs. Congress passed the Fisheries Conservation and Management Act in 1976, creating a 200-mile coastal zone. A clause in the new act let the foreign fishing continue until 1992, while American fishermen built boats to replace the fishing capacity of the foreign fleets. By the late 1990s, as I returned to school to pursue a graduate degree, it was amid a series of reports that some additional West Coast deepwater fish stocks had collapsed, this time because of American fishing pressure. The Department of Commerce declared the West Coast groundfish fishery a disaster in 2000.

If it was Gordon White and his stories about rosefish that helped send me to graduate school at the University of California, San Diego (where I had the enormous good fortune to work with Naomi Oreskes), it was Nick Bez, the immigrant millionaire fisherman and aviation pioneer, who got me hooked on trying to uncover the story of fishery development on the postwar West Coast. As part of my coursework, I arranged a work study in the archives at the Scripps Institution of Oceanography Library (a beautiful building that is, sadly, now closed). Under the guidance of archivist Deborah Day, I would archive some of the massive collection of material that the American Tuna Association had turned over to the library. There were 236 boxes, from the days when the ATA, at its offices on the dock in San Diego, One Tuna Lane, was the headquarters for development of the tuna industry in the Pacific.

It was fascinating stuff, a rich seafood broth—everything from gas receipts to lists of groceries supplied to the boats, newspaper clippings, and pictures. But mostly there were letters, reams and reams of letters. I started reading the letters for 1947. The first folder contained the "Panama Correspondence" of ATA director George Wallace from Balboa, Panama. In April of 1947, Wallace wrote that Nick Bez was in town with the boat and that the Panamanian president and his wife had been spotted on board. "Now about business," wrote Wallace. "It's intrigue, intrigue and more intrigue;

who to trust is a problem, but we are getting closer every hour to a state of confidence as to this business."[27]

ATA director Charlie Webb replied, thanking Wallace for his information about Bez and the *Pacific Explorer*. Webb ended his letter: "PS: I intend to be in Washington, D.C. on April 24 to attend the hearing that is to come off there regarding the 'Pacific Explorer.' We are hopeful that the result of that hearing may be the end of the 'Pacific Explorer' in those waters—but then again we are not sure."

Who was Nick Bez? And why would the Panamanian president and his wife be onboard a tuna fishing boat? I was still a newspaper reporter and I wanted to know more. There were further letters, about the congressional hearing, pages of testimony, and many references to Bez as "the friend of presidents." I read about the picture of the two men in the rowboat for years before I finally found a copy, at the Truman Presidential Library in Independence: Harry Truman fishing for salmon and Nick Bez rowing the boat.

This is the second book that I have written around that picture and these events. The conversation in the rowboat was certainly one the factors that led to the Truman Proclamations in 1945. The first proclamation dealt with oil and gas reserves; the second said the U.S. had the right to create conservation zones to protect fish in the waters off its coast and to restrict fishing by other nations. The U.S. would argue in vain that it had never actually created any conservation zones, but a powerful precedent had been set and countries lost no time in citing the proclamation as the authority for their own actions in banning fishing in their waters. This story is told in my first book, *All the Fish in the Sea: The Failure of Maximum Sustained Yield*. The book is primarily about the relationship between the U.S. and Peru, Chile, and Ecuador, and how this struggle shaped fisheries science. At a 1955 meeting in Rome, participants, led by the U.S., adopted Maximum Sustained Yield (MSY) as the goal of international fisheries negotiations. "The policy of the United States Government regarding fisheries in the high seas is to make possible the maximum production of food from the sea on a sustained basis year after year," wrote Wilbert McLeod Chapman, who used MSY as a tool to achieve the American policy goal of maintaining the freedom of the seas.[28] While this definition of MSY sounds imminently logical and simple, it is not, and the criticism of MSY is voluminous. Despite being amended, it is still at the heart of most international fisheries management agreements, even though its implementation, through fishing regulations that targeted older and larger fish, works against evolutionary adaptation to the environment.

The events in this book overlap with its predecessor. Both deal with the American rebuilding of the Japanese fishing fleet, the signing of the Japanese peace treaty in San Francisco in 1951, and the impact of foreign fish on domestic fishermen. I have repeated elements of the first story to make this narrative understandable. The focus is wider, exploring a little-known area of Pacific history: the postwar rivalry between the Soviets and the Americans to take over the rich fisheries the Japanese had developed throughout the Pacific during the 1930s. It looks at how trade policy with Iceland and its vitally important air bases created a loophole that let Japanese tuna into the U.S., weakening both the New England and Southern California fishing fleets. The development of efficient purse-seine technology in the 1950s, and the opening of tuna canneries in American Samoa and Puerto Rico, sent Pacific tuna boats (both Japanese and American) to develop new fisheries in the Atlantic and Indian Oceans. Events in both books took place simultaneously. I have created two intersecting stories, about the science, and about the globalization of fishing.

Most of the scholarly literature on fishery development is about the Atlantic, where fisheries developed over hundreds of years. Development in the Pacific was much faster, with fisheries developing and crashing, sometimes in the space of a decade. Since that conversation with Gordon White, I've been studying the development of fisheries and fisheries science in the Pacific, primarily through my blog, the Pacific Fishery History Project (https://carmelfinley.wordpress.com/). The blog has given voice to Charles R. (Bob) Hitz, a retired fisheries scientist who served many years aboard the West Coast's primary fishery research vessel, the *John N. Cobb*. It was Bob who traced the fate of the four trawlers built to fish for the *Pacific Explorer* and recovered the story of how the *Cobb* was built.

I have made extensive use of Bob's work, as well as three memoirs: *Living off the Pacific Ocean Floor*, by George Moskovita (1913–2000), an Oregon trawler; *The Rockfish's Warning*, by retired fisheries professor Donald Gunderson; and *The Race to the Sea*, by Dayton Lee Alverson (1924–2013), also a retried fisheries scientist. I am indebted to my friends Jergen Westrheim (1921–2012) and George Yost Harry (1919–2015), who freely shared their memories and their editorial suggestions.

My study is grounded in place and in time. It begins in Astoria, Oregon, at the mouth of the Columbia River, in 1947, when the Oregon Fish Commission hired a young World War II naval veteran, George Yost Harry Jr., to start a research program. During the war, the Army had signed contracts with fish companies, buying all the fish fishermen could catch, providing a steady market for the first time for the fledgling trawl fishery. Fishermen were moving into deeper water, finding enormous quantities of new species of rockfish that they called Rosies. The first assistant Harry hired was Jergen Westrheim;

both later worked with another young veteran, Dayton Lee Alverson. The three did early pioneering work on the bright red fish that trawlers like George Moskovita and Gordon White were delivering to the plants; Westrheim did the foundation work on determining the age of rockfish, which turned out to be far older than anyone had suspected. It was Donald Gunderson, a young fisheries scientist in the 1970s, who built on Westrheim's work as he studied the destruction of the stocks by the Soviet fleet.

The Astoria biologists may not have known Nick Bez (1895–1969), the powerful new chairman of the Columbia River Packers Association, but fisherman Moskovita certainly knew him. The powerhouse West Coast cannery was poised to enter the international tuna market. Both men were Yugoslavians: Bez arrived in the U.S. at 16 and made his way to Tacoma and its purse-seine fleet. Moskovita's father, Dome, fled conscription into the Yugoslavian Army during World War I and wound up fishing out of Bellingham, Washington. Dome and George moved to Astoria in 1939, pioneering the trawl fishery. How did this very small and localized fishery turn into a target for the massive Soviet fishing fleet?

From these five men, their ships, and their relationships, I have built a web outward, a web of connections between individuals, boats, and events, and between the documents I was finding in my archival research. Most of those papers centered around Wilbert McLeod Chapman (1910–70), the "biopolitician," as his friends liked to call him.[29] A prolific writer of letters, Chapman left extensive files with the ATA in San Diego, in the files of the U.S. State Department between 1949 and 1953, and in his papers, catalogued at the University of Washington Special Collections. The story of American involvement in the rebuilding of the Japanese fishing empire, and the impact of the tariff battle on the Southern California tuna fleet, is told in those letters.

This is a story about how Cold War policy linked fish from different oceans—Icelandic cod, Japanese tuna, and Pacific rockfish—and how the industrialization of fishing resulted in the creation of international law and the modern fisheries management system.

But most of all, this book is about rosefish, and how the Soviet fleet decimated a deepwater fish population that has not yet "recovered," despite being placed on a recovery plan in 1981.[30]

CHAPTER ONE

The Fishing Empires of the Pacific: The Americans, the Japanese, and the Soviets

> The tuna business is largely concentrated in Southern California. That industry, which always shows surface indications of bursting apart violently from internecine squabble, is capable of uniting almost instantly to give the most ruthless competition, in either the production, processing, or marketing field, to an outsider trying to come into the tuna business. . . . A newcomer to the business, without deep roots of fisheries know-how, can confidently expect to be crucified before he gets his feet under him.
>
> —Wilbert Chapman, 1949[1]

Nick Bez always said he didn't talk to President Harry Truman about the world's biggest fishing boat, but nobody believed him. After all, there was the photograph: Truman holding up a salmon that he did not catch, Washington Governor Monrad Wallgren in a sports shirt and a happy grin, the Secret Service agent in the sunglasses, and a man identified as a Seattle chef lounging on the bow of the rowboat. It is impossible to believe that Bez, in his suit, tie, and fedora, was along merely to row the boat. He was one of Washington's richest men, the owner of gold mines, salmon canneries, and the first airline to fly into Alaska. He had deep roots in Washington's Democratic Party and was on Wallgren's advisory committee on progress and industry. It was June of 1945.

Two days after the President left town, Washington Senator Warren Magnuson announced that the Defense Plant Corporation would make a $2 million loan to a new company, the Pacific Exploration Company. Nick Bez would be in charge and the new company would expand American fishing deep into the Pacific. A 423-foot World War I freighter would be converted into the world's largest fishing boat, the *Pacific Explorer*. It would stake an

Fig. 1.1. When President Harry Truman went fishing in Puget Sound in June of 1945, millionaire fisherman and aviation pioneer Nick Bez rowed the boat.

American claim on the rich Pacific fisheries developed by the Japanese: king crab and bottomfish from the waters off Alaska, and high-seas tunas in the new American possessions, the Marshall, Mariana, and Caroline Islands, deep in the equatorial Pacific.

"Tomorrow the Marianas," wrote *Pacific Fisherman* magazine.[2] No one was quite sure where the Marianas actually were, but as Magnuson said, the voyages would allow the Americans to claim expanded fishing grounds, laying a foundation for President Truman at some later date to extend the "three-mile limit to the edge of the shelf and prohibit fishing by other nationalities," the *Seattle Post Intelligencer* reported.[3]

Unspoken was that the "other nationalities" were the Japanese, the world's leading fishing nation. Their government-subsidized cannery ships and their fishing fleets prowled the Pacific until 1939. Four decades earlier, they had decisively defeated Russia, winning access to the salmon-rich waters off the eastern coast of the Asian continent. From the island of Sakhalin and the Kamchatka peninsula, it was a short hop to Alaskan waters. They sent an expedition for king crab in the eastern Bering Sea in 1929, canning crab for export to the U.S. under the Geisha brand. Americans had spent some $27 million during the 1930s buying Japanese-canned crab that had been caught off Bristol Bay, which meant that the crab had, in essence, been *American* crab.[4] By 1933, the Japanese had also developed a full-scale fishery on a medium-sized bright red fish, *Sebastes alutus*, or rosefish.[5]

American fishermen watched fretfully until 1936, when the Japanese announced a three-year experimental fishery for salmon. The politically powerful Seattle-based salmon industry pressured the State Department until it asked the Japanese government to withdraw the proposal. Japan did so, but refused to promise that it would not attempt to return to Alaskan waters in the future. Alarmed, the industry formed the Joint Committee for the Protection of the Pacific Fisheries, with Miller Freeman, publisher of *Pacific Fisherman*, as chairman.[6] Throughout the war, Freeman pelted federal officials with letters, warning of Japanese naval officers disguised as fishermen, and the danger the Japanese posed to salmon along the entire West Coast. If the Japanese were allowed to fish off Alaska, they would move on south, all the way to California, destroying the fish as they went.[7] Washington's two senators leaned on federal officials, urging them to expand the three-mile limit to the edge of the continental shelf, so that foreign boats could be prohibited from American waters.

As far as the salmon fishing industry was concerned, Truman did just that two months after his visit to Seattle, when he issued the Truman Proc lamations. There were two of them, one dealing with the jurisdictional problems between the state and federal government over newly discovered offshore oil deposits.

The second proclamation dealt with fishing. It declared that the U.S. had the right to establish conservation zones to protect fish in the high seas contiguous to the United States, where fishing activities were fully developed. If the Americans had developed a fishery, the government had the authority to limit fishing to American boats, in the interests of "conservation." In terms of international ocean law, the Truman Proclamation was an early and significant development, with a nation-state attempting to enclose a part of the oceans and to prohibit fishing by other nation-states.[8] It attempted to reserve the salmon for Americans, but it can also be seen more broadly, as the Americans challenging the Japanese, the world's dominant fishing nation, and trying to prevent them from returning to fish in Bristol Bay.

Historians have always been puzzled by the proclamation. It is so unimportant that the major historical figures who were involved, including three secretaries of state—Cordell Hull, Edward R. Stettinius, and James Byrnes—had nothing to say about it in their memoirs.[9] Scholars are extremely critical of it, calling it an "unmitigated political disaster," that the State Department would spend three decades trying to undo.[10] It flew in the face of other State Department policies of liberalizing economic and financial practices and the elimination of tariffs. The Potsdam Declaration of 1945, which set the terms for the surrender of Germany and Japan, said Japan was to have access to raw materials it had utilized before the war, and that obviously included the right to fish in the international waters off Alaska.[11]

Regardless of where the proclamation came from, it placed fish at the intersection of colliding domestic and international forces in both countries (fishing is seldom just about fish).[12] The Northwest salmon industry wanted Japanese boats banned from the waters where Americans fished. The proclamation did not explicitly ban anybody from fishing in Alaskan waters. But it went way too far for the Southern California tuna industry, which was fishing off Latin America, just as the Japanese had been doing off Alaska. Barely a month after it was issued, Mexico adopted a 200-mile exclusive zone. Argentina followed a year later, in October of 1946. Chile acted in June of 1947 and Peru six weeks later on August 1.[13] Iceland acted in 1948 and Costa Rica in 1949. The proclamation was an opportunity for many nations unhappy with foreign boats fishing and whaling in their waters and they quickly took advantage of it.

There is a large body of scholarly work that looks at the Truman Proclamation in terms of the development of international law. Much less examined is its impact on the development of fishing in the Pacific Ocean. The conversation in the row boat marked a resolution by the American government and its fishing industry to stake a claim on the fisheries developed by the Japanese during the 1930s off Alaska, as well as in the eastern tropical Pacific.

The U.S. wanted to expand its presence in the Pacific. The Joint Chiefs of Staff in a 1946 report identified 20 foreign locations where the military wanted air transit rights. The State Department had a list of bases it deemed "essential" or "required" for national security (including Iceland). Admiral Chester W. Nimitz reasoned that "the ultimate security of the United States depends in major part on our ability to control the Pacific Ocean," and he joined Truman administration officials in supporting American control over the Pacific Islands.[14] Not all the bases were built. But policy makers and the public were in agreement that America's postwar security hinged on control of the island world of Micronesia, especially the Marshall, Caroline, and Mariana Islands.[15] A line of American fishing boats would help cement the American claim to the islands, and to the vast fish resources of the Pacific Ocean, especially high-seas tunas.

The Soviets were also taking over the Japanese fisheries in its waters, including the canneries and the processing equipment the Japanese had abandoned when they left Sakhalin, the large island some 26 miles north of Hokkaido and east of Russia. If they were to build a submarine base at Sakhalin, the Micronesian Islands were the logical place for a counterattack.[16]

There was enormous enthusiasm for expanding American fishing into the former Japanese fishing areas. The Americans had fished in Alaskan waters since the 1880s, but only for salmon and halibut. Processing companies were primarily based out of Seattle, but also operated out of Astoria and San Francisco, and they sent their canning crews north for a two-month fishery. Bristol Bay, on the easternmost arm of the Bering Sea, is home of the world's largest runs of sockeye salmon (*Oncorhynchus nerka*). Dozens of lakes lance through the landscape, creating conditions where sockeye thrive. When the brief season ended, the companies returned to the Lower 48, loaded with silver tin cans of bright red and pink salmon. But the Americans had not attempted to explore any other fish resources in the deeper waters, as the Japanese (and even the Soviets) had done. There had been very little oceanographic exploration by Americans during the 1930s.[17] Until the end of the war, there were no research vessels or any funding for deep-sea research.

Investment in the Bristol Bay fisheries was estimated at $98 million, with profits of $3 million annually.[18] The 1936 catch was the largest in the history of the industry: a record pack of 8.4 million cases was produced, nearly a million more than the previous record pack of 1934.[19] Biologists had argued since 1919 that the salmon were being overfished, but that was obviously nonsense, since the catch kept setting new records as more and more fishermen entered the fishery.[20]

Congress passed a special appropriation in 1940 to investigate the development of an American king crab fishery in Alaska. The cannery vessel *Tondeleyo* and three catcher boats found large quantities of king crab, as well as bottomfish.[21] Under the leadership of Leroy Christie, a U.S. Fish and Wildlife economist, the voyages of the 113-foot boat provided evidence there was an "outstanding opportunity for a large-scale king crab enterprise in Bering Sea, and probably only a large-scale operation could be successfully conducted there."[22] But the American fishing industry was small and unorganized, incapable of investing in a big new fishery.

The largest domestic fishery was for California sardines. The catch fluctuated wildly, which scientists interpreted to mean the stocks were overfished; the sardine industry also thought this was nonsense, since record catches were still being caught as more boats entered the fishery. New England, where fishing predated settlement, was the second-largest fishery and it was admittedly an industry in trouble.[23] Boats were small and old, self-financed, and operating with limited equipment and technology. Canadian fish undercut the market for American fish. A few boats had experimented with freezing their catch, but refrigeration was not widespread, especially in smaller boats.

Fishermen were poor. Fishing was for those who could not find more attractive work on shore, as historian Samuel Morison documented in his 1941 study of the maritime industry in New England.[24] It was difficult and dangerous. Fish prices were low and markets were sporadic. Cod was caught by men in small boats, pulling lines by hand, throughout New England and the North Sea, just as they had done for centuries. Some boats were trawling with nets, but the heavy cotton nets were difficult and dangerous to handle at sea.

The coming of war would be the catalyst for enormous changes in the fishing industry throughout the world. Government money would ease the transition between salted, dried, and canned fish, to a world of frozen fish and new fish products, such as fish sticks. Government money would fund research on new technologies to catch and process fish, catching them on a massive scale previously unimaginable. Government money would also shape research and development from a haphazard process into a system of new technologies, efficiently transforming natural resources into wealth.[25] Most of all, the government money would ensure the rapid transfer of new technologies, not only to catch, but to process and market fish. All of these actions would have a major impact on fish populations.

With the coming of war in December of 1941, the American fishing industry found itself with a much higher profile. Securing the nation's food supply was urgent. Americans had been providing food to Great Britain, the Soviet Union, and other allies since 1939, but now it had to make sure there would be enough food for American troops, and that there would not be famine at home if the war was prolonged. Fishermen were on the front lines in one of the most important domestic efforts of the war, exempt from military service. When food rationing for sugar and coffee was introduced in 1942, fishermen were essential workers, exempt from complying.

The Roosevelt government moved swiftly to organize the country for war, bringing an unprecedented level of government control of the domestic economy. The War Production Board decided how steel would be allocated. The War Manpower Commission allocated labor. Food shipments had to meet the requirements of the Office of Defense Transportation. Much of the reorganization came through the Department of the Interior, run by Harold L. Ickes (1874–1952), Secretary of the Interior between 1933 and 1946. At one point during the war, Ickes held 16 separate administrative positions, included Petroleum Coordinator for National Defense.[26] He was also the director of the Office of Fisheries Coordinator, created in 1943 and charged with deciding where fishing would be allowed. In a message to the fishing industry in January of 1943, he urged maximum production for the war effort.[27] Fish had a role to play in securing victory and so did fishermen.

The federal government appropriated about 700 fishing boats during the war, placing them with the navy, coast guard, and army. Shipyards were fully occupied building military vessels, under the direction of the Controlled Materials Plan, which decided where supplies and labor would be best utilized. Protein was desperately needed and the industry was in the difficult position of being asked to catch more fish, while it had fewer boats and fishermen to do the job.

Ickes took over two of the country's largest fisheries. The Coordinated Pilchard Production Plan in California began in March of 1943, and the Controlled Production Plan in Alaska three months later.[28] Both ventures had been very successful; it was obvious that fishing was one area of the domestic economy that would benefit from some federal intervention.

Speaking with a meeting of industry representatives on February 2, 1944, Ira Gabrielson, director of the U.S. Fish and Wildlife Service, urged the industry to get better organized and to work as a unit, rather than competing with each other. If the industry was more efficient that would solve the problems of fluctuating markets.[29]

Japanese factory ships were able "to scoop up thousands of tons of wealth within a few miles of our shores," Gabrielson told the industry representatives. "Little is being done to utilize these vast resources for American industry."[30] Congress ordered another survey in 1944, of American fishing resources—including those high-seas resources expected to fall under American control at the end of the war.[31]

Working closely with Ickes was the War Food Administration, which had its roots in the Department of Agriculture, in charge of food policy and planning. Early in the war, the new agency came up with the idea of increasing the fish supply by having the federal government go fishing—but fishing on a massive scale. The ship would have to be big enough to catch and can crab, salt cod, and fillet and freeze soles and other flounders. "With a refrigerated cannery, fish could be delivered at once to the front-line supply bases in the Pacific—crab canned, the cod salted and the sole filleted and frozen," the *Seattle Times* enthused.[32] A ship big enough to do the job would cost $2 million.

Washington Senators Warren Magnuson and Hugh Mitchell pushed the proposal. It would lay claim to Pacific waters for American fishermen. The U.S. Fish and Wildlife Service was also enthusiastic, especially its deputy director Leroy Christie, who was now attached to the Office of the

Coordinator of Fisheries as head of the newly organized market development section.[33]

A World War I military vessel, the 423-foot *Mormacrey*, was available and big enough to be converted into a cannery ship that could be used to supply food to the far-flung American military bases in the Pacific.[34] The Office of Economic Warfare had already sent a team of American scientists to the Western Pacific to search for fish to feed American troops; they found tuna in the waters around the Pacific Islands.[35] The idea was soon expanded to having the boat fish itself, operated by the U.S. Army. The army was busy with other things and not especially interested in running a fishing boat. The obvious civilian choice to run the operation was the swashbuckling fisherman millionaire from Seattle, Nick Bez. He had experience running mothership operations; he was also a prominent Democrat with strong ties to both Magnuson and Monrad Wallgren, the former Washington senator who was now governor of Washington.

By the time Magnuson announced the plans, the project had grown to include the building of five large steel trawlers that would catch and deliver crab and fish to the *Mormacrey*, now renamed the *Pacific Explorer*, for processing; plans were eventually scaled back to four trawlers. A biologist was added to the project to do research on fish stocks, while a fisheries technologist would work to refine canning and freezing techniques. All the information would be freely published. There was also a new political objective: a trip to the central Pacific, to the Pacific Islands, to fish for tuna.[36] By the time the conversion was finished in a Bellingham shipyard, the cost was $3.75 million. The four steel purse seiners pushed the price tag to $4.75 million. For the U.S. government, which had spent virtually nothing on the domestic fishing industry for decades, it was a substantial investment.

No one doubted it would be a success. Magnuson's announcement was welcome news in Seattle and Alaska, but also in California, especially in San Francisco, where an ichthyologist named Wilbert McLeod Chapman was actively campaigning for government action to expand American fishing deep into the Pacific. Chapman, who would become one of the chief architects of American postwar fisheries policy, was one of the few people who actually knew where the Mariana Islands were.[37] He had spent 18 months in the central Pacific, part of the federal military expedition to scout for fish to feed American troops.

The fish had moved faster than his slow, underpowered boat with its finicky engine. Despite the crude equipment, he and his crew caught an average of 400 pounds of tuna a day. A modern American boat—like the tuna trawlers off the West Coast where Chapman had grown up, with refrigeration and a more powerful engine—would be able to catch between one and two thousand pounds a day, "for I never saw albacore off the Washington coast in abundance which approached the daily condition at Midway."[38]

From his desk, first at the California Academy of Sciences in San Francisco, then from the School of Fisheries at the University of Washington, Chapman skillfully and effectively promoted the expansion of American fishing, especially for tuna. He was a scientist but he had a passionate interest in the fishing industry and a deep well of knowledge about it. Many scientists criticize his scientific output, but no one questions his political skills and his ability to promote his views. In April of 1945, he wrote a three-page letter that he sent to all the West Coast governors, state and federal representatives, industry representatives, and the press. The copy in Magnuson's files is underlined, with pencil notes in the margins.

"From New Caledonia up through the New Hebrides and Solomon's to Green Island and back to Guadalcanal I traveled by small fishing craft, trolling all the way," Chapman wrote.[39] He laid out a thrilling vision. Developing an American fishery deep in the Pacific would require island bases, such as the bases the Americans were planning for the new island territories they would control. But "the ground work must be laid now, while our bargaining power in world politics is at its apex. Foresight now will secure to us a rich resource in the future."[40]

The military had requisitioned the largest American fishing boats, but by April of 1943 the War Shipping Administration began to return boats to fishermen.[41] Six months later, Ickes announced that 361 new boats would be built, replacing the boats lost, destroyed, or withdrawn from fishing. Prewar, the fleet had 5,600 vessels of five tons or more, and some 65,000 smaller craft.[42] The following July, Ickes approved $3.5 million of controlled materials for use in fishing boats and processing plants, including carbon steel for engines, and 1.8 million feet of lumber.[43]

It was time for the American fishing industry to modernize. The traditional fisheries—salmon, halibut, and sardines on the West Coast, and cod on the East Coast—would probably not sustain much new growth. But there were new species to find and exploit. Most important of these were high-seas tuna. Nobody knew how vast the tuna resource was, but Chapman compared them to the buffalo on the Great Plains, with fishing expanding America's Manifest Destiny into the high seas. American fishermen would replace the Japanese who had been harvesting king crab in Bristol Bay. And they would take over the longline fishery the Japanese had developed in the Marshall, Mariana, and Caroline Islands.

JAPAN

Fishing has always been of enormous importance to Japan. It was embedded into national policy, the educational system, and into village life, to a degree found nowhere else in the world. Fishing has been a vehicle of imperialism for Japan, going back to the seventeenth century. During the two centuries when Japan held itself aloof from the West, it was occupied with a brisk regional trade and a highly developed series of fisheries throughout the region.[44]

When Admiral Perry and his black ships arrived in 1854 and demanded that Japan open itself to trade with the West, the Japanese realized they could not defeat Western military technology. The government was forced to sign a series of unequal treaties that sharply limited how the economy could expand. They had to allow Western goods into the country and they could not organize themselves to compete with the imports. One of the few niches open to them was their vast knowledge of the sea. After the Meiji Restoration in 1869, the government began a series of subsidies to encourage the development of shipbuilding and fisheries. Their expansion into the high seas began in 1893, hunting for North Pacific fur seals.[45] They began to import European trawl technology in 1908 and the boats and gear were quickly copied.

"For oceanic navigation, the new Meiji state in 1868 began with nothing—no ships, no technology, no money, and no trained people," wrote maritime historian J. C. Perry. "Japan experienced a long, slow start with little achievement until the turn of the century. Until then, Japan had no steel or engineering industries, and the first modern shipyards in Japan were obliged to fashion their own steel plates or buy them abroad."[46]

Maritime power was indispensable to Japan's progress as a modern nation.[47] Japan had used junks in its network of regional trading, but with the start of the war with Russia in 1895, she needed large, modern, European-style steamers. Subsidies to the shipbuilding industry began in 1896, with bounties covering shipbuilding and the development of navigational technology. There were three phases of support: construction bounties, general navigation bounties, and special subsidies to individual lines on specified routes that provided quick and regular communications with regions where Japan had political and economic interests. By 1910, there were subsidized services to San Francisco, Seattle, and Tacoma. Between 1914 and 1922, the merchant fleet doubled, making Japan the third leading shipping power in the world.[48] An additional bounty was adopted in 1917 for the steel industry. By 1918, Japanese shipyard output had reached 489,924 tons.[49]

Shipbuilding was dominated by the Mitsubishi group and its yards at Nagasaki. Nippon Yusen Kaisha (NYK), founded in 1893, would ultimately become the world's largest shipping enterprise.[50] The Japanese military was also involved in building a navy, including the world's largest battleship, the *Satsuma*.[51]

Japan's initial focus for trade was the North Pacific. After the consolidation of the island of Hokkaido into the Empire, Japan continued its northward expansion with the Russo-Japanese Treaty of 1875. The treaty had a property swop: Russia exchanged the Kurile Islands that stretch from Hokkaido to Kamchatka, for Sakhalin, the large island some 26 miles north of Hokkaido and east of Russia. The agreement secured the northern border and marked Japan's first negotiation as equals with a Western power.[52]

The agreement allowed Japanese fishermen to continue to fish off Sakhalin, which is at the center of Asian salmon production, with enormous runs of pink (*Oncorhynchus gorbuscha*) and chum (*O. keta*) salmon. The smaller and most abundant of the five species of Pacific salmon, pinks spawn in the large river valleys of the west coast of Kamchatka and in the short, shallow rivers of Sakhalin Island and the Kuriles. There are also large runs of chums that range from northern Korea to the Arctic coast of Siberia. The largest Asian chum runs are on the west coast of Kamchatka, the north coast of the Okhotsk Sea, and the Amur River.[53]

The rapid Westernization of the Japanese economy soon led to imperial ambitions. Between 1895 and 1941, Japan greatly expanded its empire. Japan decisively won its war against China in 1895 and took its first colony, the island of Formosa (now Taiwan). Access to the salmon-rich waters off Sakhalin and Kamchatka had been one of the reasons Japan went to war against Russia in 1905. The victory gave Japan a foothold in Manchuria and later Korea. Toward the end of World War I, Japan declared war on Germany, then claimed the German colonies in the Pacific, specifically the Marshall, Caroline, and Mariana Islands. Japan took over Manchuria in 1931, invaded China in 1937, and attacked Pearl Harbor in 1941.

By 1900, government resources poured into fisheries, funding the introduction of Western trawl technology, bounties to build seagoing vessels, rebates on the salt excise duty, rebates on the import duty on oil for canned goods, closed periods for fishing, and protective measures for fishermen, including a program of weather warnings.[54] Additional bills funded the construction and improvement of fishing ports, installation of refrigeration equipment, and increased exportation of marine products.[55]

Japan led the world in developing canning technology on the high seas. Central to the Japanese system of fishery development was an extensive educational component devoted to all facets of fishing. Village schools taught calisthenics aimed at producing strong fishermen who could haul large tuna, caught on slender bamboo poles with hooks, into the boat. There was an extensive network of marine experiment stations, where Japanese scientists gathered data.

By the early 1930s, Japan was the leading fishing nation in the world. They held a virtual monopoly on all aspects of fishing in Southeast Asia, from catching to marketing, supplying 40 percent of the fish consumed in Malaya and half the catch in Singapore.[56] Northward, they had pushed past the island of Sakhalin into the waters off Kamchatka, and finally into the Bering Sea. As the war began, there were 1.5 million people involved in Japanese fisheries, with 364,000 boats bringing in catches valued at 400 million yen. Fishing boats had expanded into the Sea of Okhotsk, the Yellow Sea, the China Sea, the South China Sea, and the Gulf of Tonkin. The Japanese caught fish for food but it was also a source of currency, selling canned salmon in Britain and king crab in the U.S.

Exploratory fishing operations were launched in Mexico and Argentina by 1933. Other expeditions sought fish off the coast of northern Australia and into the Indian Ocean. Japan moved into Antarctic whaling in 1934, soon growing to be a dominant whaling nation as well.[57] Whale oil was also an important source of hard currency. The industrial conglomerate Nippon Suisan bought a Norwegian whaler and sent it to Antarctica. A new ship was built for the 1936 season, followed by five more by 1939.[58] Whaling quickly became an important source of foreign currency, as well as a food supply for the Japanese armed forces.[59] The 1938 harvest was a record 7,500 whales that produced some 80,000 tons of oil, sold in London and Hamburg.[60] Japan entered whaling despite the objections of Britain and Norway, which had historically dominated the industry. And while the boats of other countries at least paid lip service to recommendations that females and calves not be hunted, Japan included them in its catch.

"Japan's position on whaling was closely tied to its larger political goal of gaining respect in the world, which involved balancing a combination of muscle flexing and appealing to common interests," according to historian Kurkpatrick Dorsey.[61] Others have argued that Japan moved into whaling to provide foreign currency for the military. Writing in 1970, George L. Small argued that the pelagic fleets sent to Antarctica were owned and operated by the Nippon Suisan Kabushiki Kaisha Company, the main shareholder of which was the Manchurian Heavy Industries Corporation, the principal economic and industrial arm of the Japanese Army in Manchuria. The objective of the company, as stated in the *1941 Mainichi Yearbook*, was the acquisition of foreign currency and food supplies for the Japanese armed forces. The production of soya and other vegetable oils in Manchuria made it possible for the government to forbid the entry of Japanese-produced whale oil into the country. The oil was sold in Europe, especially the U.K., thereby acquiring for Japan hard currency for the war effort.[62]

Regardless of the motive, the expansion of both fishing and whaling were important components of Japan's industrialization and modernization. Throughout the 1930s, the Japanese fishing and whaling industries expanded rapidly, drawing complaints from neighboring countries. Their entry into the international waters off Bristol Bay, in hopes of catching a share of the salmon runs, marked an escalation of Japanese-American conflict.

In some ways, Japan, which is itself an island chain, is not so much a part of Asia as it is part of Micronesia, the thousands of islands created millions of years ago with the eruption of undersea volcanoes. A vast network of reefs and submerged ridges make up the floor of the Pacific Ocean, connecting Japan to many of the major island chains: the Marianas, Caroline, and Marshall Islands, Taiwan, and the Philippines.[63]

The history of the settlement of Micronesia is still not completely known. Many of the islands are uninhabitable, or have sea currents not favorable for navigation without an engine.[64] Most of the modern nations in the islands have very little land mass; most of their territory is water, and home to one of the world's great fish families—the tuna stocks of the Western Pacific. The islanders were sophisticated navigators and built sturdy outrigger canoes that could handle turbulent waters, but they traditionally did not engage in much interisland trade. Resources on all the islands were similar and groups were self-sufficient. The islanders were technologically primitive, widely scattered, and ill prepared to resist invaders.[65]

The one significant natural resource for the islands is tuna. Tuna are pelagic, which means they wander the open sea, mostly in the upper layers of the oceans, heated by the sun. They are biologically very advanced, streamlined for rapid swimming, and they maintain a body temperature that is higher than the water in which they swim. Tagging studies show they migrate thousands of miles across open oceans.[66] When bait fish are found, the tuna go into a feeding frenzy. The "fish behave from minute to minute, hour to hour, and day to day in response to oceanographic, atmospheric, and solar-lunar processes that operate on similar time and local space scales."[67] It took the development of boats with motors to chase tuna in the ocean.

Ferdinand Magellan's last ship landed at the Mariana Islands in 1521. Spanish armadas followed; then Dutch merchants and missionaries, American whalers, and French and British warships. The European powers were engaged in colonial expansion, finding new territories they could exploit to further their trade. Over a 300-year period, the Great Powers would control the oceans and that control helped to consolidate political power in Western Europe. The Europeans were the only nations that could operate in all the oceans at once. The maritime empires they created were on an undreamed-of scale.[68]

The invaders carried disease, severely reducing the population of the islands. The Marianas had an estimated preinvasion population of 50,000, but by 1710, there were only 1,500 islanders. When the first American ship visited Guam in 1802, the indigenous Chamorro population had virtually disappeared.[69]

The European struggles over colonial territory accelerated between the 1880s and the start of World War I. Mark R. Peattie calls this period the new imperialism, when leaders trumpeted the rise of an industrial system, with its demands for new sources of raw materials and markets.[70] There is no question that industrialization conferred benefits to native cultures in Africa, Southeast Asia, and in Micronesia. There is also no question that it literally destroyed some of the islands, made others uninhabitable, made nomads out of some islanders, and would ultimately cost a great deal of money to maintain. It also had dire implications for tuna stocks.

Japan was also industrializing and colonizing. By 1931, she had a fleet of 13 factory ships and more than 100 catcher boats working mainly in the area near Kamchatka. Around 60 percent of the catches were salted or frozen on board, while the rest was transported to land factories for canning. There were also 17 mothership operations producing 300,000 cases of king crab a year, an important source of capital.[71] They were also hunting fish to the south, sending a factory ship, the 1537-GT *Haruna Maru*, into Southeast Asian waters in 1932; the ship had packing facilities.[72] Japanese fishing operations made money, and a key component was paying the crew very little.

Takiji Kobayashi (1903–33), the most famous writer of the proletarian school in Japan in the 1920s, wrote about the inhumane treatment onboard a Kamchatka crab ship in his short story "The Factory Ship," published in 1928. It was based on a newspaper account of the brutal treatment of the crew on a crab boat in the Okhotsk Sea. In some of the versions of the translation, the ship is poaching in Russian waters, it is accompanied by a naval ship and fishing is linked to Japanese military expansion:

> To hear them tell it, Kamchatka and Sakhalin and this whole area will have to become Japanese territory someday. The place is richer than anybody dreams it is . . . Don't you believe it when they tell you that the

> destroyers are sent out only to protect the factory ships. That's not their only purpose—not by a long shot. There's a more important goal. They're making a careful survey of the sea and weather conditions in this area up to Sakhalin and near the Kuriles, getting ready for the day. This is supposed to be a secret, I think, but they're hauling big guns and oil on the sly to the last island of the Kuriles chain.[73]

Japan was also interested in the expanding to the south. In the 1880s, Japanese romantic literature began to feature the possibility of an oceanic empire. Popular writers debated whether Japan's future lay in *hokushin* (continental expansion) or *nanshin* (oceanic expansion).[74] The government sent samurai families to Western Micronesia in the 1890s, to an area the Japanese called the *Nan'yo*, the South Seas. Germany had bought the islands from Spain after the Spanish-American war in 1899. The Japanese found German merchants who were trading in copra, the dried flesh of coconuts, from a series of small islands.

With the end of World War I, Japan moved quickly to claim the islands, arguing they deserved them as their contribution to the war effort against Germany. The Americans, concerned about Japan turning the islands into strategic bases, inserted a nonfortification clause into the peace treaty.[75] The islanders were not in a position to argue for themselves. All of Micronesia was hit hard by the influenza epidemic of 1918–19. All of the islands lost people, with the highest mortality in Western Samoa, which lost 22 percent of its population in a matter of weeks.[76]

Micronesia became a Class C Mandate of the League of Nations in 1920, a status that demilitarized the islands but allowed them to be governed as an integral part of the empire of Japan. In 1921, the administration of Japanese Micronesia passed from the military to the Ministry of the Interior. Troops were sent home and an imperial decree in 1922 established a separate South Pacific administration, directly subordinate to the prime minister. They took over the phosphate mines on Angaur, in the West Carolinas, and the railroads and other facilities in the Marianas. Efforts to integrate Micronesia into the empire intensified as the 1930s went on.[77] The development of the "Co-prosperity Sphere" under the administration of the South Sea Government was well underway.[78]

The Western model of imperialism called for traders to move to the colony and send exports back to the home country. The Japanese did it differently, creating a network of industry and communications within the colony, and bringing industry to the labor and raw materials.[79] As was their pattern, the Japanese transferred expertise from Japan, Okinawa, and Korea to colonize and subsidize commercial enterprises. The Japanese also greatly improved health care, started an education system, imported modern amusements, and set up an economy where cash replaced barter as the main means of exchange.[80]

The Japanese initially argued that they occupied the islands for security reasons. They built new harbor installations, wharfs, roads, and public and private buildings, proving they were modernizing and improving the islands. They experimented with growing rice, cotton, and sugarcane, augmenting the general trade of copra. "Almost every island supports its own small experimental program," wrote historian Hermann Joseph Hiery.[81] In the Carolinas, fishing, phosphate mining, and the shell industry predominated. Several agricultural experiment stations were established in the Marianas and Carolinas, aimed at improving sugarcane and cotton production, as well as testing various vegetable crops.[82] In the middle of 1915, the Japanese calendar and time were introduced. The islands were a significant source of food and other raw materials for Japan before World War II.[83]

The Japanese were the first to commercially exploit tuna in the islands, harvesting some 300 million pounds of tuna by 1923.[84] There were almost 400 boats by 1938, catching tunas and bonita and shipping the fish to Japan for processing.[85] It was a lucrative fishery.

THE SOVIET UNION

Unlike Japan, which spent centuries developing its relationships with the sea, the Soviet Union was first and foremost a land empire. The official figure for the national fish catch in 1913 was only about 900,000 pounds, most of it landed from inland lakes and rivers, with a small portion coming from the shallow waters close to the Arctic and Pacific shores.[86]

It took until 1930 for the Soviets to catch that many fish in a single year again.[87] During the 1930s, British trawlers caught most of the fish landed in the Barents Sea. In the Soviet Far East, Japanese boats took 85 percent of the salmon, thanks to concessions wrung from the Russians after the loss of the 1904 war.[88] Throughout the 1930s, the Japanese had increasingly intercepted Kamchatka salmon on the high seas, greatly reducing the numbers coming back to the rivers to spawn.

Starting in 1945, the Soviets began to build what would become the world's largest whaling and fishing fleet, with more than 5,400 distant-water vessels and amounting to at least half the world's gross vessel tonnage for fleets of this size and type.[89] The Soviet expansion into fishing was initially enormously successful, beyond anyone's wildest dreams. A series of small coastal fisheries were turned into a global industry, fishing throughout the world's oceans. It was an expansion stopped only by the biological collapse of the fish—and the whales.

The expansion of fishing was only part of the larger drive for Soviet superiority on the seas. Soviet leaders set out to dominate the seas. They built an oceangoing fleet, a powerful navy, merchant marine, and oceanographic research vessels. Like Japan, they used this emerging maritime strength as evidence of their modernity and rapid economic growth.[90] The Soviets would challenge the military dominance of the U.S. and Japan over the rich marine wealth of the northern oceans.

Historians are slowly uncovering the environmental legacy of the Soviet Union's push toward industrialization. "They moved inevitably forward through the forest, across the steppe, and into the tundra, rarely slowly," wrote historian Paul Josephson. He might also have added that they moved deep into the world's oceans, exterminating both fish and whale species. There was never any public opposition to the wildest of plans.

"Even before one project had been completed, the ministry, construction trust, or institute responsible for it would secure approval for another massive project to ensure that the workers stayed busy, whereas at the end of a project in a market economy the workers would lose their jobs," Josephson wrote.[91] State propaganda glorified the Soviet worker and idealized his selfless commitment to socialism. Citizens were urged to expand as "rapidly as possible so the Soviet Union could become the most powerful country on earth."[92] Rapid industrialization was the key to the spread of socialism.[93]

When Stalin took over in the late 1920s, he oversaw the "Great Break" with the past, through a series of five-year plans that established targets for production to increase the supply of grain, make the land more productive, feed the population, and provide grain for export. The plans called for the rapid industrialization and collectivization of agriculture.[94] The first five-year plan forced the farmers to grow unfamiliar crops, such as cotton and sugar beets. The plan was poorly administered, leading to significant amounts of grain left unharvested, while millions starved. Half of the nation's livestock was slaughtered rather than given to the collective farms. By some calculations, by the end of the Soviet period, collective farms used three to five times more chemical pesticides, herbicides, and fertilizers than a farm in the U.S., poisoning the land and increasing the potential for erosion.[95]

The state wasted enormous resources on grandiose projects that failed on an even grander scale. Plans were initiated in Moscow, with little technical input or regard to alternatives and feasibility. There were not enough tools and machinery, and workers were often poorly housed, with inadequate food, in the inhospitable far north. Much of what Josephson and others have written about the industrialization of the Soviet Union—the waste, environmental destruction, and the suffering imposed on workers—is shared by the workers involved in the expansion of Soviet fisheries.

There was an enormous demand for electrical power. The creation of large hydroelectricity facilities led to the destruction of the fisheries on the inland lakes. The most important was the northern Caspian Sea, where the Volga River delta provided a rich environment for many kinds of fish, including sturgeon, beluga, whitefish, and various types of herring. Dam construction on the Volga severely reduced the amount of water into the sea, exposing more land mass and reducing the area available for the rearing of fish. Construction of the Volga-Moscow Canal began in 1932. Hydroelectric dams and reservoirs were built at Uglich, Ivankova, and Rybinsk. The lake shrank by 40,000 square kilometers—an area equal to the size of Belgium or Holland—disrupting not only fishing, but also shipping and trade.[96] The negative impacts on fisheries were noted as early as 1933.[97] The creation of a series of reservoirs changed the flow, current, temperature, and biological composition of the river. Engineers planned to ameliorate for the negative consequences with artificial spawning areas and fish breeding plants. They were unable to halt the population declines.

A number of factors led the Soviets to expand into fishing. They had progressively expanded the empire, annexing Western Belorussia and East Prussia. The shoreline extended for thousands of miles and there were 10 seaports, four of them accessible throughout the year.[98] Since all citizens were guaranteed employment, the shipbuilding and fisheries expansion was an important source of employment. But the real motivation was the desperate need for protein.

Fishing had not been a large part of the economy, except perhaps in coastal areas. The chaos of World War I and the Revolution destroyed whatever fisheries had been developed. The new government invested in building a northern fishing fleet. There was a strong demand for food and catches were good, leading to systematic state investment. The Murmansk herring fleet began to fish in the cold waters between the Atlantic and Arctic Oceans and developed a year-round fishery before it was disrupted by war.[99] There were also two experimental new ventures. They began whaling in 1932, with a converted American freighter, in the waters off Kamchatka. They added a second whaling ship in 1946.[100]

Between 1931 and 1935, the Pacific Institute of Marine Fisheries and Oceanography (TINRO) did a series of exploratory trawling expeditions to the Okhotsk and Bering seas. The trawler *Paltus* was dragging southeast of the Pribilof Islands and lifted a three-metric-ton catch of large rockfish, *Sebastes alutus* and *S. polyspinus*. The catch was so impressive the agency recommended starting a commercial fishery. The institute fished for three years, until the fleet found large quantities of flatfish in the southeastern part of the Bering Sea. The exploratory work on rockfish was put on hold.[101] The war ended the experimental fishing, but the Soviets would be back.

The Great Stalin Plan for the Transformation of Nature was adopted in 1948, a massive plan to reforest the steppes. Its proponents claimed no capitalist country could accomplish such a feat.[102] There were also massive irrigation projects. The Karakum Canal, started in 1956, the longest irrigation canal in the Soviet Union, stretched some 1,300 km westward into the Kara Kum Desert. It was one of the least efficient irrigation systems in the country. Its planners did not foresee large quantities of salt blowing on the farmlands, the increased potential for dust storms, and the loss of biodiversity in the lake itself.[103]

This pattern was repeated in the 1960s with the Aral Sea, with development drastically reducing the size of the world's fourth-largest freshwater lake. Between 1960 and 1987, its level dropped nearly 13 meters and its area decreased by 40 percent, due to water withdrawals for irrigation. The fishery had employed 600,000 people.[104]

The two winters following the end of World War II were extremely harsh for all of Europe. Russia had lost at least 25 million of its people. Agricultural production in 1945 was half of what it had been in 1940, which had not been an especially good year.[105] A drought in 1946, combined with the population dislocation of the war, resulted in the worst grain harvests in a century, not only in Russia, but through much of Europe. Harvests improved during 1947 and 1948, but they were still low, an irritation to a government intent on showing the world the success of the communist system.

What fishing boats the Soviet Union did have were heavily damaged during World War II, leading the government to its first five-year plan in 1946. The restoration of the fishing industry would be accomplished through "the principle of improvement and modernization."[106] Plans called for increasing harvests far beyond the prewar level; the fisheries goal for 1950 was to increase the catch by at least 50 percent to 340,000 metric tons.[107] As part of that expansion, they began a campaign of whaling in 1948.[108]

During the next two decades, fish played an important role in the Soviet economy, providing up to one-third of the nation's animal protein. It was an important source of employment. The fishery was managed by a large, centralized state administration and supported by educational establishments to provide a large workforce. The Soviet state subsidized the fishery, initially because it was exploratory, then for strategic reasons.[109] The industry enjoyed several advantages. Funds for capital construction came from the central budget and did not have to be paid back. The price of fish was set by the state and the higher the price, the easier it was to show a profit. Workers were paid by piecemeal or by a standard rate, with incentives in the form of extra pay for long service, overtime, overfulfillment of quotas, and regional rates for hardship service on long or stormy voyages.

By the end of World War II, there was a small fishery in the southern Barents Sea for herring, cod, and plaice. The herring fishermen were the first to fish further afield, moving to Spitsbergen and Iceland in 1949 and to the Greenland Sea in 1951. They found good fishing. The Soviet catch in the North Atlantic rose from 5,000 metric tons in 1949 to 261,300 metric tons by 1956. By that time the fleet had expanded to 450 medium-sized trawlers. Baltic-based ships were active in the Norwegian Sea in the early 1950s, while Murmansk-based boats were fishing in the Northwest Barents Sea.[110] To keep catches high, they continually expanded to new grounds.

There was another force working to expand the Soviet fleet, and that was the Lend-Lease Program in the U.S. Lend-Lease was intended for Great Britain, China, and the countries of the British Empire; the Soviet Union was added in 1941. Total aide exceeded $50 billion, with $31 billion to England, and $12.5 billion to the Soviets.[111] During the war 8.1 million metric tons of American ships went to Russia, and the registry reflects another 5.3 million metric tons transferred to them.[112] The program included $21 million spent on fishing boats, paying for 23 boats, including 10 cannery vessels and eight refrigerator vessels.[113] Several Soviet cargo ships were converted into floating canneries to pack king crab. Conversion of the 360-foot *Alma-Ata* was done at Northwest Iron Works in Portland, Oregon, a project that took 12 months and cost close to $2 million. When President Truman cancelled the Lend-Lease Program, a private contract with a Soviet purchasing agency provided the funds to complete the conversion. The War Shipping Administration approved building nine 42-foot boats to supply the massive canning vessel.[114]

The U.S. was not only going ahead with creating its own deep-sea canning vessel, the *Pacific Explorer*, it was paying to help its wartime ally go fishing itself. And it would soon be creating policies to restore the fishing fleet of its former enemy, Japan.

CHAPTER TWO

Islands and War

The scholar's facts, it has been said, are not fish on the carving board but fish in the sea.
—John Dower[1]

War transforms geographical relationships, sometimes overnight. Backwaters become strategic assets. Small countries are thrust onto a bloody world stage. Some countries thrive under their new geographical importance—none more so than Iceland, located on the great circle route that Allied planes would fly to the airfields of Britain and the Soviet Union. As war approached in 1939, Iceland hoped to remain neutral. But the German Navy sent a cruiser to their waters in 1939 and the airline Lufthansa applied for transit rights, which Iceland refused. Throughout April of 1940, Adolf Hitler's soldiers marched through Europe, occupying Poland, Denmark, and Norway.[2] Would they try to take Denmark's Atlantic colony, Iceland?[3] It was vital that Iceland not fall to German troops.

Hitler launched his blitzkrieg against Belgium and Holland on May 10, 1940. The same day, British and Canadian troops invaded Iceland and took control of its capital, Reykjavik. They dropped leaflets apologizing for the "disturbance," and hoping it would not last long. The Icelandic government made a formal protest against the violation of its neutrality and independence. Then it urged its citizens to treat the invaders as guests, displaying the greatest courtesy.[4]

The early part of the war did not go well for the Allies. The Germans occupied France, and by early in 1941 had attacked Yugoslavia, Greece, and North Africa. In June of 1941, the German troops began Operation Barbarossa, their march into Russia. They routed British troops in Greece and brought them to a standstill in North Africa. The troops in Iceland were needed elsewhere. The U.S. had not yet entered the war, but within a month, the two countries concluded a defense agreement that called for American troops to occupy Iceland for the duration of the war. The Americans promised they would support Iceland's claim of independence from Denmark, keep Iceland supplied with goods, and make "favorable commercial and trade agreements with it."[5] They also promised to withdraw their troops "immediately on the conclusion of the present war."[6]

The Americans arrived on July 7, 1941. They rapidly built a major naval base at Hvalfjörður, and the Keflavik airfield at Reykjavik became one of the largest wartime bases—crucially important for antisubmarine activities and protection for the North Atlantic convoys that were supplying the Allies. Iceland's ports were ice-free all year, ensuring that manpower and supplies from the U.S. would reach England to stage for the liberation of France.

With a population of 80,000, Iceland was one of the poorest countries in Europe.[7] It was transformed by the war. It went from an almost feudal economy, struggling to sell salted and dried fish in southern Europe, to a strategically important nation, with new demands for its fish and labor. The war brought wealth and modernization to what had been a remote island. Iceland escaped from an old set of geographical limitations through a new set of relationships, turning the traditional tilt away from Europe and toward America.[8] The British and American troops stimulated the local economy, bringing jobs and hard currency. But many thought the bases invited attack from Germany, making the country less safe.

Iceland had always been occupied: it was first settled by the Norwegians in the tenth century.[9] Denmark assumed control around 1262 and there was a long period of harsh exploitation, with Danish kings and merchants bent on extracting as much profit as possible from their colony.[10] A Danish treaty with Britain in 1490 gave British ships fishing rights off Iceland. The first British steam trawlers appeared in Icelandic waters in 1890; by 1904, there were 180 trawlers reported in Icelandic waters, more than three-quarters of them British.[11]

Iceland's waters are home to some 300 species of fish; the most important have traditionally been cod and herring. The continental shelf is wide, extending over 150 km in some areas, and is cut by many subsea canyons. Beyond the shelf, the sea floor falls away from 200 to 400 meters on the shelf to a depth of more than 1,000 meters. Three major current systems influence fish production: the warm and saline Irminger current, an offshoot of the Gulf Stream flowing from the south; the very cold and less saline East

Greenland current from the northwest; and the intermediate East Icelandic current from the northeast. The mixing of warm and cold currents stimulated the production of plankton and other marine organisms.[12]

When Europeans introduced steam engines into fishing boats during the 1880s, the new engines greatly increased their ability to fish further from home. Steam engines freed fishermen from the capriciousness of the wind and the strength of their bodies to haul in their nets. Engines allowed them to use larger nets and to fish in deeper water. "With each improvement in fishing equipment, such as the change from sail to steam, from steam to gasoline motors, and from gasoline to distillate to diesel fuel, the intensity of fishing increased and the fishery was pushed farther and farther away from the ports of landing."[13]

Icelandic historian Jon Th Thor placed the expansion of fishing within the context of British imperialism: "The trawlerman sailed hither equipped with the most modern fishing gear known at that time and were backed by the most powerful naval force in the world. On the coast of Iceland they met people who for all practical purposes were in great contrast to themselves, people who hardly kept alive under the most primitive conditions, many surviving near starvation-limits and still remaining at stage of industry befitting the middle Ages. Off their coast these people possessed some of the world's richest fishing grounds, but they had neither the funds nor the equipment to utilize them."[14] The depletion of local stocks in the North Sea and North Atlantic was so marked that in 1902, the North Sea governments created the first organization devoted to fisheries science, the International Council for the Exploration of the Seas (ICES) based in Copenhagen. Boats sought new waters to fish. Iceland was the extreme range of steam trawlers and the crew loaded as much coal as possible into every nook for the journey. Until larger vessels were built, Iceland was a summer fishery, because the winter fogs made the voyage so hazardous.[15]

As Danish power waned by the nineteenth century, Iceland pressed for a relaxation on the restrictions the Danes had imposed on its economy. The Althing, Iceland's parliament, was reestablished in 1843 and Iceland began a long campaign for independence. The Union Agreement was signed in 1918 and would run for 25 years. After 1940, either nation could revise or cancel it.[16]

Fish have always been at the heart of Iceland's relationships with the world. Icelandic fishermen had scant capital to build larger boats or fish processing facilities, and the government could do little to police fishing in its waters. During the 1920s and 1930s, fishermen in rowboats caught much of the cod, just as they had for centuries. The fish was salted and dried for sale in

Portugal, Italy, Spain, Brazil, and Cuba, although some salted fish also went to Britain, France, and Argentina. Most of the herring catch was cured for sale to Sweden and only small amounts of fish were frozen.

The government tried to be aggressive in defending its fishermen from the impact of the foreign boats. An 1889 act forbade trawling within near shore waters, but it could not be enforced. A stronger act was passed in 1894, banning vessels with trawls from entering local waters or putting in at Icelandic ports, except in emergencies. The British trawlers argued this was contrary to international law, but several vessels were arrested, fined, and had their gear confiscated for illegally fishing off Iceland and the Faeroes.[17]

The government made another attempt to deal with overfishing in 1937, this time trying to use science as a way to conserve stocks. It persuaded ICES scientists in Copenhagen that Faxa Bay, an important nursery area on the western coast, was being hit hard by foreign trawling. ICES suggested holding an international conference to discuss a scientific experiment, a 10-year moratorium on fishing. Britain declined to participate and the conference was never held.[18]

The concept of a three-mile limit was commonly accepted, but it had never been put into law. The League of Nations held a conference in 1930 to see if there was any consensus on how the growing number of fishery disputes could be handled. Iceland argued hard for a wider territorial sea, to force foreign boats out of its coastal waters. The U.S., Britain, and Japan were opposed to any regulation. With no agreement, boats from a dozen nations continued to fish in Icelandic waters, producing an estimated 17 to 21 percent of total European fish production by the start of the war.[19]

Iceland fumed at its inability to police its own waters and protect its main asset. It pushed ahead with efforts to develop its industry. With the outbreak of the Great Depression, markets in southern Europe deteriorated. The Spanish Civil War disrupted the importation of fish from Iceland. Britain and Germany imposed quotas on fish imports. As demand fell, so did prices.[20] There was widespread unemployment, a factor in the rise of the Icelandic Communist Party, which controlled 10 of the 52 seats in parliament in 1942, the peak of its political power.[21] The Communists were especially unhappy about the air base and the presence of the Americans.

As promised in the defense agreement, the U.S. signed a favorable trade agreement with Iceland in 1943. It lowered the tariffs on a number of items, including fish canned in anything except oil. The normal U.S. tariff rate on fish canned in oil was 45 percent; the new trade agreement set a tariff of 12.5 percent on fish canned, for example, in water, or perhaps sauce. It was a minor clause and it is doubtful if anybody paid much attention at the time.

After all, fish was canned in oil. There was no reason to can it in anything else.

The government built its first experimental frozen fish plant in 1929; within two decades, Iceland had 62 freezing plants.[22] The government also created a community around the herring processing plant at Skagastrond, on the north coast.[23] The most significant step in support of the industry was the passage of Act No. 34, creating a Fisheries Fund, capitalized at 100 million kroner. The fund was an independent institution under the supervision of the minister in charge of fisheries and managed by the Fisheries Bank of Iceland.[24]

As the war deepened, the German Navy began to blockade British ports. The British fishing fleet—already decimated by the navy claiming its larger boats—was kept in port. There was a desperate need for protein, and Icelandic fishermen willing to brave the long and dangerous run to Britain received premium prices for their catches. Through the war, Icelandic boats supplied 70 percent of the fish consumed in Britain. Fishermen caught a record 880 million pounds of fish in 1944, with 512 million pounds of it exported.[25]

With the end of the war on June 6, 1945, Iceland wanted the American occupation to end. Having spent centuries freeing themselves from Danish sovereignty, they did not want to wind up an American colony. While the fighting was over for the Americans, the air base was more vital than ever. They sought to make their presence more favorable to the Icelandic people. Iceland wanted preferential status for its imports to the U.S., a "status that went against the whole thrust of American policy of free trade."[26] If the Departments of State and Defense wanted to keep the air base, they would have to find a way for Americans to buy more Icelandic fish.

THE PACIFIC ISLANDS

While Iceland was able to leverage its strategic geographical importance and its access to fish stocks to strengthen its economy, other islands were not so lucky. The Marshall, Mariana, and Caroline Islands in the equatorial Pacific were successively invaded by German, Japanese, and American colonizers.[27] Their geographic importance was due to their remoteness; they were a base within the vastness of the Pacific for trade between the U.S. and Asia, as Iceland was on the way to Europe. The remoteness allowed nations to move into nuclear testing. While Iceland only tilted toward the West, maintaining its autonomy, the Pacific Islands were engulfed by their new strategic importance. Some of them were literally destroyed, as part of the American

Cold War strategy. For the islands, as historian Kimie Hara has argued, the Cold War has not yet ended.[28]

Displacing the Japanese in the islands had been a long and extremely brutal endeavor. U.S. marines fought to destroy extensive Japanese fortifications in some of the deadliest fighting in the war.[29] Slightly more than a quarter of the American casualties during the war (107,000 out of 407,000 dead) occurred during the battles of the Marshall, Mariana, Caroline, Volcano, and Ryukyu campaigns. The Americans had paid "blood and treasure" for the islands, as Fleet Admiral Nimitz put it in 1946.[30]

With the end of the war, the U.S. planned to absorb the new territory into its Pacific sphere of control. They had tried to prevent the Soviet Union from obtaining occupation rights to the Kuriles, the chain of islands that stretch from Hokkaido to Kamchatka, separating the Sea of Okhotsk from the Pacific Ocean, and to Sakhalin. The Japanese had controlled the island since the 1905 war with Russia. According to the Yalta agreement, Sakhalin was returned to the Soviets. Just as the Americans sought to take over the Japanese fisheries in Alaska and the Trust Islands, so the Soviets were hoping to replace the Japanese with their own fishermen in the Soviet Far East.

Many military leaders believed the U.S. had been unprepared for the war in the Pacific and must never be caught in that position again.[31] The unilateral Soviet control of the Kurile-Sakhalin area was perceived as a strategic naval threat from the Soviet submarine fleet in the Pacific, according to historian Hal Freidman. The Soviet Union "might be able to complement this submarine threat by building a surface navy and strategic air force in the future. To American naval officers, the best way to prevent this Soviet strategic threat was to take direct control of Micronesia, using the islands as part of a strategic basing system for deep strikes into Soviet territory in the event of war and a deterrent in time of peace."[32] As early as 1943, army and navy planners resolved to establish a defense in depth by taking permanent control of the Mandated Islands acquired by Japan after the First War.[33]

The American troops who went ashore on the islands after the fighting ended described broken communities, where natives were living in caves and suffering from hunger, thirst, and lack of medical treatment. The American bombing killed many, destroyed the railroads and sugar mills the Japanese had built, as well as many homes. The destruction was so great survivors could not identify where their homes had stood.[34] The islands were littered with leftover munitions and the economy had been set back by at least a quarter of a century.[35]

The Americans saw little of value in the islands, except that their remoteness and small population made them ideal for nuclear testing. Having wrested control of the Pacific from the Japanese, they did not want to lose it to the Soviets.[36] They insisted on absolute American control. The islands needed to be modernized, to play a role in resisting the expansion of communism in the Pacific. While other nations after World War II dismantled their colonial empires and allowed new nations to move toward self-rule, the Americans delayed that process, especially in Oceania.[37]

The Americans portrayed themselves as motivated by the well-being of the islanders, not as exploiters of the islands' resources. The American objective was to turn the islands from "international liabilities into assets," Francis B. Sayre, the U.S. representative on the United Nations Trusteeship Council, wrote frankly in 1948.[38] If the islands possessed assets, they must be utilized for the common good of the United Nations. "The former Japanese mandated islands are of little or no economic importance," wrote Sayre, but he added that they were of enormous strategic value, which far outweighed any social and commercial considerations.[39]

The end of the war left the fate of the islands unsettled. The United States Commercial Company, a subsidiary of the Reconstruction Finance Corporation (RFC), began an intensive effort to rebuild the economy, promoting American-style agriculture, handicraft production, and development of trade. Anxious to find out about its new responsibilities, the U.S. Navy launched the Coordinated Investigation of Micronesian Anthropology. American social scientists were flooding into the islands in what was perhaps the largest single social science project ever undertaken outside the U.S. There were 42 anthropologists, linguists, and geographers from 23 universities and museums in what one anthropologist called a "massive ethnographic salvage program."[40]

The social scientists were "insisting that we must build an economy for these small areas and that social reformation of the natives must take place," E. C. Weitzell, an agricultural economist with the U.S. Department of Agriculture, wrote in 1946.[41] "The native 'way of life' is not to be modified readily. However, constant drive and encouragement will produce reasonable results."[42]

The Americans rejected the language and the concept of colonialism, but they had long wanted Pacific resources for themselves. The resource they had wanted most was the magnificent harbor at Pago Pago, on the island of Samoa, some 2000 miles southeast of the Marshall, Mariana, and Caroline Islands. American Navy Officer Charles Wilkes visited the island in 1839 and described the harbor, formed by the encircling walls of a submerged volcano, with a narrow opening and harbor protected from all storms. Tutuila is about 2,500 miles north of the Australian city of Brisbane. It was just 750 miles northeast of Fiji, and about 2,500 miles southeast of Hawaii. The harbor was big enough to shelter the entire American fleet—at least at first.

The fleets of other nations also visited Pago Pago. The French arrived in 1786; British Methodists came about 1830 and set about converting the natives; American Protestant missionaries began working with the islanders in 1857.[43] The Germans also established a trading house in 1857, buying copra.[44]

All of this jostling over Samoa intensified during the 1870s. The Americans set up a naval coaling station and repair base at Pago Pago in 1878. The Germans were on the island, but the Americans were only interested in the harbor. After decades of local skirmishing and Washington-Berlin diplomacy, the island was finally divided in 1900. The Germans controlled the western portion, while the Americans controlled the eastern portion, known as American Samoa.[45]

A naval officer was dispatched to secure an agreement to establish the naval station. "It was incidental that the naval station and its surrounds happened to be inhabited," wrote I. C. Campbell. "The U.S. thereby acquired responsibility for about 5,700 Samoans without any commensurate interest in their future, usefulness or welfare."[46] Over the next three decades, a succession of navy officers would oversee the island.

The military had set up some fishing and fish processing to supply food for civilian internees and military personnel, but the program did not work out well. There were three sets of processing facilities and they were mostly unused. "It is extremely doubtful whether private interests in the United States will be concerned with commercial fishing in these distant waters," wrote Weitzell.[47] The fishing was more productive closer home, where processing plants were established.

THE UNITED STATES

When George Moskovita brought the *Treo* to Astoria, Oregon, in 1940, he was 26 years old and had been making a hardscrabble living as a fisherman for a decade. He was the son of immigrants, his mother from Ireland, his father from the Adriatic, where he had fled to avoid conscription during World War I. George had been a deckhand, purse seining for salmon in Alaska, sardines off California, and tuna off Mexico. He'd scraped the money together to buy the 55-foot *Treo*, an old wooden boat with a gas engine. There were no controls in the pilot house, just a bell to signal the crewman in the engine room to put the engine in or out of gear. It's likely that he just intended to set crab pots near the mouth of the Columbia River, dangerous enough for an old boat designed to purse seine in the sheltered waters of Puget Sound, not to drag nets in the open ocean.

"They were paying a dollar a dozen for crabs at first, but when they chopped the price to fifty cents a dozen, we quit," Moskovita wrote in his memoir, *Living Off the Pacific Ocean Floor.* "We shopped around to see if we could sell some drag fish because we also had our drag nets with us. But nobody wanted to buy bottom fish. We were the only ones with dragging gear in Astoria, so the packers didn't handle it."[48] The only market for trawl-caught fish was the mink farmers and they paid less than two cents a pound.

Eventually one of the fish plants offered three cents, although Moskovita and his crew had to unload the fish themselves. "We went out about four in the morning and came back about seven or eight in the evening to unload," Moskovita wrote. "Then we'd go again the next day if the weather was good."

Two months earlier, the *Treo*'s engine burned out its bearings on the way to the fishing grounds and had to be towed back to port.[49] Now the *Treo* hit bad weather on its way back to Astoria and started to take on water. Moskovita drained an inch of gasoline from the engine into a bucket, went to the stern, and lit it with a match. "It was raining and blowing, but the flame coming out of the bucket made a pretty good flare. You could see it for a long way. And, by golly, a boat was coming. It was the *Washington*, skippered by Captain John Lampi, out of Ilwaco."

Moskovita and his crew had been bailing with hand pumps and buckets. Lampi tied a towline and started to tow the *Treo* into port. The towline broke and the boat was filling up with water. It was midnight, cold, raining, and blowing. Eventually all they could see was the *Washington*'s mast light, disappearing in the distance.

> I knew we were going down so I got out the life preservers. I put my arms through the straps in one, but it fell to pieces and dropped to the deck. They were old and rotten. We should have had better survival gear, but even if we would have realized the condition of what we had, we didn't have any money to buy any. So we had to use what we had. So, it was looking pretty grim on the deck of the Treo. Then all of a sudden, we saw red and green lights. The Washington was coming toward us again. He got to us and threw a line which all three of us on board grabbed a hold of at the same time and we jumped overboard. By the time he pulled us over to his boat and got us on board, the Treo had gone under.[50]

Fig. 2.1. The *Treo* sank one December night in 1939, almost costing George Moskovita and his crew their lives.

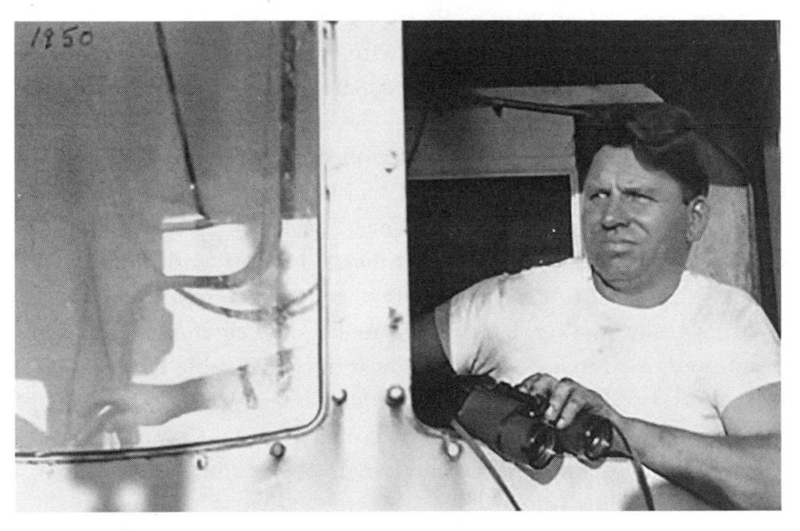

Fig. 2.2. George Moskovita brought the *Treo* to Astoria in 1939, one of the first boats to trawl for groundfish.

The story of the sinking boat made the front page of the *Daily Astorian* the next day. "I lost everything when that boat went down," Moskovita wrote.[51]

Fishing was on the cusp of an enormous transformation. The coming of war would be the catalyst for a new relationship between the fishing industry and the nation-state, not just in the U.S., but globally, in ways that would have a major impact on fish stocks. By the 1940s, the war in Europe had rippled out and touched American fishermen, even small-boat fishermen such as George Moskovita in Astoria, Oregon. After Germany invaded Norway on April 9, 1940, the Nazis diverted all Norwegian food items back to Germany, including all of the production of cod liver oil. Vitamins had not yet been synthesized and the oil was a rare source of vitamins A and D.

The shortage of high-value vitamins was an opportunity for the West Coast fishing industry. They went fishing for sharks. Most of the oil came from the livers of dogfish shark (*Squalus suckleyi*) and soupfin shark (*Galeorhinus zyopterus*). Buyers paid a dollar per pound for the oil, and sometimes much more. Moskovita describes a winter fishery during the war, where he made about $25,000. "We were getting $8 a pound for the liver and every fish had five pounds of liver," Moskovita wrote. "That's about forty bucks for each shark."[52]

One time Moskovita made a delivery and one company offered $14.40 a pound for shark oil. "I almost fell off my chair. I couldn't believe it. I said I would be right down. I couldn't get there fast enough. They said they had tested them three times. We had about 30 pounds of livers, so the two boats split the money and we were really happy."[53]

Oregon landings peaked in 1943, with 270,000 pounds of shark landed. The next year, landings sank to 50,000. Scientists learned to synthesize vitamins and the market disappeared.[54] The Oregon fishery ended in 1949.[55] The high prices had stimulated fishing off the entire West Coast. War was transforming the fishing industry. It had been dominated by the salmon fishery. Now the season was much longer, the shark fishing and new otter trawl fishing both went for nine months of the year, with four months of albacore fishing, starting in July. There were nine canneries in Astoria.

With a ready market, boats could make large catches. "They pick up the enormous load to one side of the boat at considerable hazard while the craft reels to port or starboard, often in ticklish seas," wrote the *Oregonian*. "These boats have been known to sneak in over the Columbia River bar after four days of fishing with more than 90,000 pounds of fish dipping the rails and decks below water."[56]

By August of 1944, the fishery service was advising that the fisheries of the Pacific Northwest and Alaska offered opportunities for small businesses, with several fish species abundant enough to support large fisheries.[57] Since West Coast populations were less dense than on the East Coast, most of the catch (sardines and salmon) had been canned, and only a small percentage marketed as fresh or frozen. The increase in population due to wartime expansion meant there was enough population to support a different—and more modern—fishery, one dependent on freezing, creating new products that would ease the workload for modern housewives.

People were flocking to the Pacific Northwest. Hydroelectric units completed in 1943 at Hanford in eastern Washington supplied the abundant, cheap power needed to attract industry. Many companies that built plants during World War II stayed for postwar manufacturing and new industries moved in. The federal government's wartime role boosted state and local economies in many ways. There was money to expand port facilities and improve harbors. There were wartime contracts, such as the commission to the Columbia River Packers Association in Astoria, Oregon, to build six large wooden barges for the U.S. Maritime Commission.[58] After the war, the cannery opened its own shipyard on the site.

Trawling, using nets to fish for the fish on the ocean floor, had been slow to come to the Northwest corner of the United States. It had spread northward from San Francisco after the 1880s, but the early ventures were sporadic; many of the boats sank or the processors failed.[59] Now the army bought everything fishermen could catch, especially a medium-sized, bright red rockfish that fishermen called Rosies, or rosefish. It could be filleted and frozen and it seemed to be available in large quantities.[60] The federal government contracts played an essential role in helping the industry move into these new markets.

There was little money for systematic exploration of the seas until after World War II, when state fishery agencies became more systematic about studying fish. They hired research staff and began to publish their work; the Oregon Fish Commission published the first edition of its research briefs in April of 1948. They set up a groundfish research division, based out of Astoria, and found a young naval officer with a master's degree in zoology from the University of Michigan in 1941. George Yost Harry Jr. hired two young biologists, Jergen Westrheim and Dayton Lee Alverson, and together they started the systematic study of the red rockfish that Oregon trawlers were landing in increasing numbers.

Decades later, Alverson wrote about the work in his memoir. He made a trip out of Astoria onboard the trawler *Harold A.* He was not sure the

Fig. 2.3. The 80-foot *Yaquina* was built in 1945, one of the first West Coast fishing boats designed to fish in deeper water. With Captain Gordon White, the *Yaquina* pioneered fishing in deeper water for rosefish. Gunderson Marine permission.

boat would make it over the Columbia River bar, but it did and the skipper headed deep into the Pacific, deeper than fishermen usually fished, to 105 fathoms of waters. The first haul was astounding.

"There had to be at least twenty or more species in the checkers, ranged in length from ten to twenty-six inches," Alverson wrote. "The skipper, who was helping me sort fish, said to me, 'Lee, what the hell species are these? I am fishing deeper than we normally fish and we don't usually catch these species of rockfish. I was told by some fishermen down in Newport, Oregon that they had been picking up a small species of red snapper in the deep water and they were being sold as Pacific Ocean perch.' "[61]

Alverson got out Clemens and Wilby's *Fishes of the West Coast of Canada*, and set about identifying fish. There were 27 separate species, six that he could not identify, and others listed as uncommon or rare. Most were long-jawed rockfish (*Sebastes alutus*). They were the same fish the Soviet fishermen had found in such numbers in their 1931 expeditions to the Pribilof Islands in the Bering Sea.[62] It is likely that the bright red fish made up a substantial portion of the groundfish Japanese boats had taken from the waters around Bristol Bay in the 1930s.

The *Harold A* delivered its fish in Newport, most likely to the Yaquina Bay Fish Company, run by Dudley W. Turnacliff and Harold Penter. Alverson spins a good yarn about drinking with the fishermen at the Abbey bar on the Bayfront (alas, now demolished). They caught 42,000 pounds of fish; the rockfish sold for four cents a pound and the flounders for five. The gross was just over $2,000. The boat share was 37 percent, leaving $1,300 for the crew to share (minus food and fuel). It was, by the standards of coastal fishing standards at the time, a very good trip. And it augured well for further good trips in the future. With the end of the war, there was enormous optimism about the potential for the American fishing industry to expand, especially on the West Coast.

Turnacliff and Penter found a group of investors and raised $100,000 to build a state-of-the art trawler, the *Yaquina*, one of the first nonmilitary ships built at the Gunderson Brothers Engineering Corporation in Portland. Launched in February of 1945, it was 80 feet long and equipped with a mechanical refrigeration system that would hold fish for a month at sea. It could trawl with a net or it could jig for tuna.[63] And it could also fish further offshore, setting a net in deeper water. Gordon White started as a crewman on the *Yaquina*, but he was soon running the boat and landing big loads of rosefish. There had been a fishery on them off California during the war; now large quantities were being found in deeper water off Oregon.

The Yaquina Bay Fish Company filleted the fish and sold them into the fresh fish market. Turnacliff thought the fish were similar to east coast perch and started to label them "ocean perch." Markets increased slowly.[64] The new fishery suffered a hiccup when the army, which had taken most of the rockfish caught during the war, cancelled the contract. Rosefish were plentiful, but there was less meat on them than on other marine species, such as soles; they were more expensive to process. People didn't know what rockfish was or how to cook it. *Pacific Fisherman* bemoaned the crisis in May 1946, warning about how imports of fish had increased. "There was a good deal of talk about congressional action and to increase duties, embargo imports, and all sorts of strenuous economic actions," wrote *Pacific Fisherman*. "In general, informed persons believed there was little chance of such actions prevailing."[65]

Despite the rockfish marketing problems, fishing on the West Coast boomed after World War II. The steady markets during the war had allowed some fishermen to invest in newer, bigger boats. The U.S. Navy had appropriated many of the larger tuna vessels in California and purse seiners from Alaska. Now, with the war winding down, new, smaller ships were offered to the industry. During early 1944, the navy gave permission for a series of new tuna boats to be built and 17 vessels entered the fishery in 1945. Another 30 YP-class (Patrol Class) boats were ordered by the navy, and once the war ended, they were offered to fishermen. By 1946, *Pacific Fisherman* reported that American fishing capacity had been restored to prewar levels. And capacity—the ability to catch fish—was increasing rapidly.

The new boats had bigger, more reliable engines, allowing them to operate further offshore, and making it practical to purse seine in the ocean. Fishermen installed brine refrigeration systems, allowing tuna to be held for a longer period of time. Recording echo sounders allowed fishermen to identify schools of fish in the water, so they could set nets more precisely. Sonar searched the waters ahead of them, while Lorans made it possible to return to the same patch of water where fish had been found. While the new technology greatly expanded the capacity of fishermen, it was not immediately adopted everywhere.

The federal government authorized a round of building fishing boats, to help shipyards wean themselves off military contracts, but the boats were sent to Germany and the Soviet Union. There was no other systematic program to help the domestic industry. Some smaller ex-military vessels were sold to California tuna fishermen after the war, stimulating the growth of the tuna industry. But in Washington and Alaska, the industry was attracted to much larger military vessels, two and three hundred feet long, that they wanted to modify into freezer ships to transport salmon and tuna back to the Lower 48 for canning.

Americans had to "take intelligently the leadership in world aquatic industries that seems about to be thrust into its hands," Secretary of the Interior Harold Ickes told Congress in February of 1945.[66] The federal government had never adequately taken care of the nation's fish resources. Now, with fisheries about to expand in Alaska and the equatorial Pacific, legislation was needed to expand the "economic, technological, and biological aspects of the development, utilization, and maintenance of the fishery resources." The report to Congress was accompanied by a letter from Ira N. Gabrielson, Director of the Fish and Wildlife Service, who wrote that conserving fish "is an empty and outmoded philosophy which can have little place in the economy of a progressive people."[67] In order for resources to be worth conserving, they must first be useful.

The report estimated the total capitalization of the fishery resources at $5.8 billion, and pointed out that commercial fisheries, taken as a whole, were among the least progressive industries in the country.[68] Fishermen were slow to improve their products, develop new resources, and increase fishing efficiency. The industry was composed of small, independent users, concerned generally with local conditions and bedeviled by unstable supply and markets. That would have to change if the U.S. was to take advantage of the new opportunities the war had brought to the American fishing industry.

Unlike other food industries, the fishing sector had little assistance from the government. States controlled their waters to three miles and there was no authority to manage fisheries beyond that. Congress had never given the Fish and Wildlife Service the authority to carry out a unified national program. The report concluded with four recommendations: (1) that federal spending on fisheries be increased to levels that were comparable to other food industries; (2) that Congress authorize the Fish and Wildlife Service (FWS) to organize "a dynamic program that anticipates the needs of the Nation's fishery resources as a whole," (3) that Congress authorize FWS to actively cooperate in scientific researches with other countries that share fishery resources; (4) that federal legislation be enacted to control water pollution and to coordinate national water projects.

Rep. Joseph R. Farrington of Hawaii introduced a bill to provide for the investigation and development of fisheries in Hawaii in 1945. Japanese fishermen had supplied tuna to the Hawaiian market. Now American boats would have to supply the fish and they would need research on where and how to fish. The bill sought $350,000 for onshore facilities, $500,000 for a full-scale oceanographic research vessel, and $175,000 for an experimental fishing vessel, budget levels that were astonishingly high compared to the tiny amounts that had been spent on ocean research during the 1930s.[69]

The bill was championed by Wilbert Chapman, from his desk at the California Academy of Sciences. Throughout the next two years, Chapman worked to mobilize West Coast fishing interests to support expansion of research efforts in the Pacific. The Japanese would be back throughout the Pacific in no time. The Soviets had sent scientific fishing expeditions to Antarctica and into the eastern Pacific the previous summer. "Both these nations are moving more alertly than we are," Chapman wrote. "In the tropical Pacific we have won an empire of tremendous size. It is an empire of great riches, where the land is as nothing and the sea is everything—an empire in which the native people are small in number and restricted to small points in its vastness; an empire which no other nation save the Japanese covets and which no other nation save theirs and ours can cultivate and make produce."[70]

He wrote to Seattle's Nick Bez, asking him to bring the bill to the attention of President Truman. "It is quite possible that our fishermen could be shut out of potentially rich fishing areas in the Pacific in the future under the terms of our own Proclamation," Chapman wrote to Bez, who was busy overseeing the reconstruction of the *Pacific Explorer*. "We have strong potential competition in the area from Japan and Russia. Both of these nations are moving more alertly than we are."[71] Bez undoubtedly threw his support behind Chapman's efforts to ensure that American boats could follow the *Pacific Explorer* into the central Pacific.

It was against this framework of manifest destiny and excitement that the federal agencies had to make a decision about what to do with the *Pacific Explorer*. Bids were never issued and from the start there were questions about the seven-year contract, which called for the federal government to pay to convert the ship and build the four fishing vessels that would fish for it. Bez was to lease the ship and its four smaller boats and pay the RFC $50,000 a year, or 55 percent of the profits, whichever was greater. Bez was responsible for all operating expenses and information from the voyages was to be made public for other fishermen to use. It was an unusual financial arrangement, and it placed the federal government in the West Coast fishing business.

By the time the war ended, $2 million had been spent on the *Pacific Explorer* project. The RFC asked the Office of War Mobilization and Reconversion what it wanted to do. In turn, they asked the Departments of State, Navy, and Interior. All three advised completing the project, citing the presidential proclamation issued after Truman's trip to the Northwest, asserting the need "to conserve and develop all fishing grounds contiguous to the nation's border."[72]

In a 1944 joint resolution, Congress had directed the U.S. FWS to survey the extent and condition of all marine and freshwater fishery resources, including the "high seas resources in which the U.S. may have interests or rights."[73] Surely that would include laying claim to vast swaths of the Pacific, establishing American fisheries that could deny entry to the fishing boats of other countries?

The *Pacific Explorer* was almost twice the length of the first factory processing ship, the *Fairfree,* under development in Scotland by the whaling firm Christian Salvesen & Company.[74] It built on the seven decades of experience and knowledge that the American fishing industry had acquired about canning salmon, both off the West Coast and off Alaska, home of the world's richest salmon run, the sockeye of Bristol Bay. But all the people who conceptualized the project, including Nick Bez, knew very little about tuna. And tuna are nothing like salmon.

Rowing the presidential rowboat made Bez a national figure. He was born Nikola Bezmalinovic, on August 25, 1895, on the island of Brac in the Adriatic Sea, the oldest of six children. He learned to fish as a child. When he was 15, he borrowed $50 from his father and booked passage to New York, where he worked in a Brooklyn restaurant for the train fare west, where there was a community of Dalmatians in Tacoma, Washington.[75] He shipped as a deckhand on a towboat bound for Alaska, and when he returned to Seattle, he acquired a rowboat, then a gas boat; by 1914, he had his first seine boat. In 1922, he went to work as a superintendent at an Alaskan cannery, saving his money until he heard about an abandoned cannery on Peril Strait, near Sitka. He paid $5,000 for the plant and went into debt to buy $150,000 worth of canning machinery. The first year he paid the bills. The second year he netted $175,000.[76] In 1931, he bought three single-motored Lockheed Vegas, recruited a cadre of bush pilots, and a few weeks later, Alaska Southern Airways was making the first scheduled flights in the history of the Alaska territory. Bez sold out to Pan American three years later and plowed the profits back into the salmon business.

By 1939, Bez owned three of the largest canneries in Alaska and two gold mines.[77] He also owned the Alaska Southern Packing Company, which operated a floating cannery called *La Merced.* He created a new company, the Intercoastal Packing Company, and bought a 390-foot steamer called the *Orgontz,* which he intended to convert to a floating cannery. The *Orgontz* was equipped with two Continental Can Co. high-speed machines, two machines that gutted and took the heads off the fish, replacing the need for Chinese labor, and four retorts, fishing with a company of gillnet boats in Bristol Bay.[78] Bez certainly had big plans; he hoped to also have boats fish halibut for delivery to his motherships.[79]

Bez designed much of the *Explorer* conversion himself. It would operate with a plane to scout for tuna. The fish would be frozen and transferred back to Astoria for canning. The boat would do everything, including use tangle nets to catch crab, and process tuna and groundfish caught by its fleet of four (originally five) fishing boats.

Seattle naval architect H. C. Hanson designed the 100-foot steel-hulled clipper-trawlers, which are better known now as Pacific Coast combination vessels. The first two were built in Astoria during 1946, first the *Oregon,* then the *Washington.* The next two were built at Long Beach, California, the following year, *Alaska* followed by the *California.*[80]

The contracts were controversial. In a 1945 article, *Business Week* questioned why Bez was singled out to be given the lease and reported that the War Foods Administration and the RFC were shifting the responsibility for naming Bez to head the project. "In Seattle, Bez angrily gave his own version. Washington, he pointed out, had thought up the floating cannery project and urged him to take charge, as lessee, although he had suggested that he manage the operation without pay under army supervision. As to continuing the project in peacetime, the money involved was 'small potatoes' compared with other RFC enterprises."[81]

Bez entered into an agreement with the Transamerica Corporation of San Francisco, the investment affiliate of Amadeo P. Giannini's California banking kingdom, in 1946. He bought the Columbia River Packers Association, one of the West Coast's largest processing companies. He also bought two ex-military vessels—both 326-foot Landing Ship Tanks, or LSTs—and a military landing craft, and converted them into refrigerated motherships, each capable of carrying 1,000 tons of frozen yellowtail from the South Pacific back to Astoria for processing.[82] The vessels were renamed the *Tinian* and the *Saipan.*

With the *Pacific Explorer,* the *Tinian,* and the *Saipan,* the Columbia River Packers Association, already a West Coast and Alaskan powerhouse, was the flagship enterprise for the American fishing industry to stake its claim on the fish resources of the Pacific. Perhaps, if the *Explorer* had been delivered on time, it might have all worked out. But the conversion took longer than expected and when it was finally delivered to Bez in late 1946, he decided to make the shakedown cruise, not to the icy Bering Sea, but to some warm tropical waters to catch some tuna. After a final outfitting in Astoria, the *Pacific Explorer,* accompanied by 12 large trawlers rigged for purse-seine fishing, set off on in January of 1947 for Costa Rica.

CHAPTER THREE

Manifest Destiny and Fishing

Yet, of all the kinds of fish which support great fisheries, it is hard to find one about which we know less than we do about tunas.
—Pacific Oceanic Fisheries Investigations, October 1949[1]

World War II was a profitable time for Icelandic fishermen, especially those brave enough to run their fish into British ports, where fresh fish fetched a premium price. The British fishing fleet had been decimated when the navy claimed its largest vessels for the war and German blockades kept their boats in port. Throughout the six long years of war, Icelandic fishermen supplied about three-quarters of the fish consumed in Britain. They caught a record 880 million pounds of fish in 1944, and 512 million pounds of it was exported.[2] The 1930s problems of selling dried and salted fish appeared to have been solved.

With the war winding down, the government moved to modernize the fleet. It had already passed Act No. 34, creating a Fisheries Fund, capitalized at 100 million kroner.[3] It went shopping for fishing boats. It ordered 45 new boats from Sweden and proposed to buy an additional 50 boats in 1945.[4] Thirty new trawlers from Great Britain began to arrive in 1947; they cost $500,000 each.[5] There were 732 vessels in the Icelandic fleet by 1947.[6] In the space of a couple of decades, fishermen had gone from longlining in dories to setting records for the most fish caught in a single trip, 380 tons in just seven days, during 1948.[7] The trawler *Ingólfur Arnarson* (named after the first settler in Iceland) was the first fishing boat in the world with radar. By 1950, almost the entire trawler fleet was equipped with radar and automatic sounders and it was considered, at that time, the most modern trawler fleet in the world. The government money meant all the boats modernized at once, with great rapidity.

"From the first use of new equipment, techniques or approaches, only a few years pass until the entire fleet is following suit," wrote scientist H. P. Valtýsson, in a reconstruction of the Icelandic fishery for the Fisheries Center at the University of British Columbia.[8] The new technology greatly increased the ability of fishermen to find and catch fish. They could map the grounds and return to the best spots; their echo sounders could follow fish on their migration. The herring fleet, so important to the national economy, was at the forefront of the use of new technology. The first Icelandic purse seiners experimented with sonar in 1954. They were early adopters of the hydraulic power block, created by San Diego tuna man Mario Puretic in 1955. The block made it easier and safer to haul heavy nets from the sea. Boats started installing the block in 1959 and within two years, the whole purse-seine fleet had the equipment, almost completely replacing traditional drift netting.

The wartime inflation left the kroner overvalued compared to the economies in its traditional markets. To make matters worse, the northern herring fishery—which had been so reliably profitable during the war, providing much of the fish sold to Britain and thus providing the operating capital for the government—almost failed in 1945. Catches were poor again in the three following years, causing severe financial and political problems. This set of circumstances put fish at the heart of the government's financial and political planning, and thus at the center of Iceland's Cold War relationship with the Americans, the Soviets, and the British.

The 1941 defense agreement between Iceland and the U.S. specified the Americans would depart, "bag and baggage, upon the cessation of the emergency."[9] Nothing had been said about a continued American presence after the actual fighting ended. By June of 1945, Iceland saw the war as being over. The Americans saw the air base as still being central to its defense. Troops were still stationed in Germany, Britain, France, Italy, and Africa.[10] Withdrawing from Iceland would set an undesirable precedent. The Americans wanted a lease that would allow them to stay, but their presence was still controversial and many Icelanders argued that the air base made them a target. Icelanders wanted the troops to go home as soon as possible.

The Communist Party, which held two of the six posts in Iceland's cabinet, was vehemently opposed to any agreement that allowed the Americans to stay. The air base agreement was so controversial that the Communists were expelled from the government and a new government was formed, one more friendly to American interests.[11] The bulk of the American troops left by June 15, 1946, and a new agreement was drawn up to allow the air base to continue operations, paid for by the Americans.

It was during these tense negotiations that the Soviet Union offered to buy all of Iceland's 1947 fish production. Britain, Germany, and Norway were rebuilding their fishing industries and supplying their own markets with fresh fish, cutting into Iceland's markets and worsening its balance of trade. The Soviet Union and its satellites, Poland and Czechoslovakia, wanted to buy fish. The Soviets had not bought fish from Iceland since 1936, but now, a decade later, they wanted to resume a trading relationship.[12]

The American Embassy thought the Soviet offer was politically motivated. If the fish didn't sell, the government could fall, opening the door to a new government with two of the six cabinet posts potentially held by the Communists—possibly even the aviation ministry charged with negotiating the airport lease! The embassy thought the Russians were stalling over the negotiations, counting on the deteriorating economic situation to bring down the government. In a secret telegram, it urged the State Department to find a market for the fish.[13]

"It can be assumed that a principal aim of Communist cabinet ministers would be to do everything in their power to embarrass the United States and to nullify the special rights we have under the Airport Agreement," wrote Hugh S. Cumming Jr., chief of the State Department's Division of Northern European Affairs.[14] He urged the British to do what they could to buy more Icelandic fish. The War Department came up with $1 million to buy Icelandic fish for the relief program in the Mediterranean countries.[15] It offered to buy between 15,000 and 20,000 tons of wet salted fish, but would only pay 13.5 cents a pound. The government had guaranteed its fishermen 17 cents a pound.[16] Despite heavy lobbying from the embassy about the political importance of the deal to Iceland, it fell through.

A new airport agreement was signed on October 7, 1946, granting the Americans military air transit rights for six and a half years. As soon as the agreement was signed, the Soviets withdrew their offer to buy all the fish. Iceland's exports for 1946 amounted to $46 million, with a third of the production going to Britain, a quarter to the Soviet Union, and 14 percent to the US.[17]

Iceland was once again having problems selling its fish. Its trade deficit was growing and the government didn't want to devalue the kroner. Instead, it implemented an emergency measure to guarantee minimum prices for various goods, including fish. If the fish sold on the world market for less than the guaranteed price, the government would make up the difference.

The emergency measure was implemented in 1946 and then incorporated into an economic bill passed the next year. The government guaranteed to make up the difference if fish sold for less than the minimum price.

The price supports cost the government $3.5 million in 1947 and would become a major problem. Export subsidies grew from 30 million kroner in 1951 to 247 million in 1956, a measure of the disparity between Iceland's production costs and the export prices at which Iceland had to sell its fish and agricultural products to compete on the world market.[18]

As the Americans saw it, the government could have fallen if the fish had not been sold. That would have created an opportunity for the Communists to capture more seats in the government, jeopardize Iceland's relationship with the U.S., and potentially open the door to amend or cancel the Keflavik agreement. The Reykjavik Embassy pressed the State Department to sell the fish in Germany.[19] The embassy pushed again in a May 2, 1947, telegram, saying it was in the best interests of Americans to assist the Icelandic government in disposing of the fish—and not selling fish to the Soviets.[20] Compounding the problem was steep inflation in Iceland, at least partly because of American military spending during the war.[21] To make matters worse, the all-important late herring season got off to an unexpectedly slow start. The government had borrowed 1.5 million British pounds that it hoped to repay after the herring oil sold on the export market.[22] Meanwhile, the local Communists continued to argue that the government could solve its problems simply by exporting more fish to Eastern Europe.[23]

The Americans queried the government about countermeasures to prevent the Communists from taking over. They wanted definite plans to combat a communist coup.[24] The embassy drew up a contingency plan to deploy American forces to Iceland "to support a democratic system of government against internal Communist subversion." The wording was later changed to "protect U.S. and North Atlantic security interests in Iceland in the event of an emergency."[25] The Americans had wanted a stronger agreement over the air base that would have allowed its use as a future military base but the Icelanders refused any expansion, much to the embassy's dismay.[26]

In February of 1948, Iceland entered into trade negotiations with its two largest customers, the British and the Soviets. Both countries agreed to each take 40 percent of Iceland's production of herring oil, and to purchase two pounds of frozen fillets for each three pounds of oil.[27] The U.S. was instrumental in arranging an agreement with the British to buy 70 tons of fish for "Bizonia," as the merger between the U.S. and British occupation zones in Germany was called, announced on May 29, 1947.[28] The British were reluctant, and it took "insistent prodding" by the U.S. to get the fish agreement signed in London on December 13, 1947. Iceland hoped for a similar agreement for the 1949 catch. "It would be in the interest of the United States Government to help Iceland dispose of her fish elsewhere than in the Union of Soviet Socialist Republics, in order to prevent the Soviet Union from gaining any influence in Iceland. The United States War Department has a vital interest in Icelandic air-rights, since Iceland is strategically located."[29]

Why was it so difficult for Iceland to sell its fish? An American study showed it cost Iceland 15.3 cents to produce frozen fillets, while American fillets cost 24 cents to produce. Fish from Iceland entered the U.S. with a 2.5 cent tariff per pound, which was reduced to 1.78 cents a pound after 15 million pounds. The Americans wanted Iceland to market its own fish within the U.S.[30] Iceland wanted a governmental bilateral agreement and preferential treatment for its exports.[31] That went against the thrust of American economic policy in dismantling tariffs and creating open markets. The Iceland reciprocal trade agreement was an example of redeeming a foreign policy commitment, rather than promoting beneficial commerce. The agreement made little sense, according to Alfred E. Eckes Jr., the historian and former trade commissioner. Norway and Canada were the major suppliers of items like herring oil, not Iceland.[32]

Iceland was unable to maintain its economy without outside help. Assistance came from American Secretary of State George Marshall on June 5, 1947. The Marshall Plan was not directed "against any country or doctrine but against hunger, poverty, despair and chaos." Initial plans called for the Soviets to participate, but they withdrew from a planning meeting in Paris in early July.[33]

Iceland signed the Marshall Plan on July 3, 1949. The U.S. had already given $600,000 in Marshall Plan aid during 1948, along with a $1.9 million loan. Other loans followed, from U.S. foreign aid agencies, the International Bank for Reconstruction and Development, and the European Payments union. The U.S. loaned Iceland nearly $20 million after 1948. An additional $73 million was spent at Keflavik air base, to upgrade its facilities for use as a major air base by the newly formed North Atlantic Treaty Organization (NATO).[34] Iceland was one of the charter members.

With the American money, Iceland drew up a four-year plan to build herring liquefaction plants, a fertilizer plant, several fish meal factories, and several refrigeration plants. It would also expand its fishing fleet and increase fish production by 34 percent, from 552,000 metric tons in 1947, to 738,000 metric tons by 1952. Iceland also used an additional loan of $2.3 million to buy a fish processing factory ship, the *Haeringer*, which arrived in Reykjavik in October of 1948.[35] A new herring plant for Reykjavik using the latest technology was planned for 1950.

The infusion of money did not solve the problems with the fishery, or the rest of the Icelandic economy. The government was forced to devalue the kroner by 42.5 percent in March of 1950, in hopes of making its fish exports more competitive on the world market. The policy of guaranteeing minimum prices for fresh fish and subsidizing the export price of frozen fish was leading to turmoil in the fishing industry and that had the potential to collapse the entire economy.[36]

And it did not solve the problem of foreign boats. British, German, and Norwegian boats were once again fishing in Icelandic waters, and catching the crucial northern herring stocks that had been declining for the previous six years.[37] Attempts toward international regulation of fishing had failed and Iceland argued that its principle responsibility was to protect its fish and regulate foreign activity in its waters. It claimed a 50-mile limit in 1948, using the precedent of the Truman Proclamation of 1945, the same precedent used by the Latin American allies unhappy about American tuna boats in their waters. The U.S. refused to recognize or acknowledge the claims from Latin America. It would be more difficult to ignore the demands from Iceland.

JAPAN

With the war finally over and the surrender of Japan underway, General Douglas MacArthur, head of the Supreme Commander Allied Powers (SCAP), wanted to get as much of the fishing fleet back to work as quickly as possible. People were starving. The larger boats had been conscripted for the military during the war, just as had happened in other countries. The Allied carpet-bombing campaign had destroyed much of the infrastructure of the ports and harbors, as well as many fishing boats. But there had been an estimated 364,000 boats in Japan by 1939, and there were still many left—small boats, perhaps operated by sail, but capable of catching fish if gasoline and supplies were made available to them.

The Americans had started planning to occupy Japan in 1942 and they intended to rebuild the Japanese economy along American goals. They placed constraints on the economy, just as the unequal treaties had during the 1870s. Japan was not allowed to make aircraft, synthetic oil, or rubber. There were limits on their steel, chemical, and machine tool industries. Much of the country's food supply had been brought in from the satellite countries the empire had occupied. "Now the Japanese no longer have the fisheries of the Kuriles, the pulp of Sakhalin, the sugar and rice of Formosa, or the rich minerals and agricultural products of Korea and Manchuria," wrote historian Edwin Reischauer, recognizing the extent of Japanese imperialism.

"These areas are no longer economic tributaries to Japan, and the Japanese cannot exploit them for their own benefit."[38]

Just as it had in the 1880s, Japan turned to the sea to rebuild. Prewar, Japan had been the third-largest shipping nation in the world, carrying two-thirds of its trade in its own vessels.[39] Its fishing fleet led the world in landings, and fishing expanded substantially during the 1930s throughout the Pacific and into the Indian Ocean. Japan joined the world's whaling fleet in 1935, buying a Norwegian whaler, then replicating it and quickly becoming a leading whaling nation, arousing much resentment in the process.

Fishing had always been subsidized by the Japanese government. It encouraged the building of larger fishing boats, engaged in an extensive series of support services for fishing, including setting up a system of weather warnings, investigations in oceanography, gear technology, and the training of young fishermen.[40] It invested heavily in refrigeration technology in the 1930s, building freezing facilities and installing refrigeration on its larger fishing boats and transport vessels.[41]

SCAP built on these policies when it established priorities to allocate scarce material and fuel for the construction of new fishing boats. During the six years of the American occupation, SCAP was enormously successful in rebuilding the world's largest fishing fleet—a fleet that for decades had been too large for Japan's home waters. SCAP authorized building 795 steel fishing boats in 1946, a huge boost to the economy, despite the low wages paid to industrial workers.[42]

The fleet was rebuilt by June of 1947, with far greater capacity than before the war.[43] The restoration of the fleet was so successful that SCAP announced it would not approve any further large-scale boat construction programs. However, it established a more limited program to replace unserviceable and obsolete vessels.[44] Once fishing boat production was established, it was difficult to interrupt.

The Japanese fishing fleet had once roamed the Pacific and voyaged into the Indian, Atlantic, and Antarctic Oceans. Now fishing was tightly constrained to a small area around Japan's four islands, called the MacArthur Zone. The Japanese government and its highly organized and integrated fishing and whaling companies immediately began to push for access to the stocks they had fished before the war. Meanwhile, the fishing pressure grew on the local stocks, especially in the East China Sea.

According to a SCAP press release, the Japanese government and the fishing industry carried on a "systematic campaign to break out of the restrictive fishing area assigned to the Nipponese since their surrender and to resume their exploitation of distant waters."[45] The release cited SCAP's fishery director as saying that "some of the requests were for areas so far from the home islands that there could be no doubt the Japanese were aiming to lever their way back into the world market." They certainly were. There were few other choices.

Whaling was authorized in the vicinity of the Bonin and Kasan Islands in November of 1945. A far more controversial SCAP decision came on August 6, 1946, when whaling was expanded to include Antarctic waters, infuriating the British and Norwegian governments and their whaling industries. In November of 1947, two factory ships, 12 "killer boats," and seven carrier boats left Japan for Antarctica.[46] By December, a first installment of 900 tons of whale oil was sent to Bremen for distribution in the British and American occupied zones. The first shipment of whale meat and blubber arrived in Japan in January 1948.[47] A second expedition, in November of 1947, was larger and carried observers from SCAP, the United Kingdom, Norway, France, and Australia.[48] Nations officially disapproved of Japanese fishing, but they were also anxious to learn what they could for transfer their own fishermen.

The industry was anxious to expand and so was SCAP. Under the Potsdam Declaration, Japan was entitled to have access to resources it had used before the war; those resources obviously included fish and whales. SCAP's Diplomatic Section was also interested in using fisheries to advance its foreign policy objectives, especially toward the Soviet Union, which had claimed a 12-mile exclusive limit that the U.S. was refusing to recognize. The Diplomatic Section wanted to expand Japanese fishing in the area north and northeast of Hokkaido, challenging the Soviets who wanted to establish their own fisheries in the waters the Japanese had controlled. As always, there was more involved than just fish.

The Japanese fleet was confined to its home water and quickly exhausted local stocks, prompting boats to fish out of the zone. There was no money for government patrol boats. Japanese fishing boats were being seized in Korean, Chinese, and Soviet waters, drawing complaints.[49] The continued seizures fueled the American perception that the Japanese refused to obey fishing regulations. If the Japanese refused to obey the SCAP fishing boundaries, what would happen when a peace treaty was signed and there were no restrictions on where they could fish? The Americans deplored the violations of the zone while approving of needling the Soviet Union. "There is a strong possibility that the USSR will attempt to use the present fishing boundary in the north as a basis for a permanent political boundary . . . it is highly desirable at this time that it be made a matter of record that SCAP does not consider this northern fishing boundary a fixed and permanent line."[50] The

Diplomatic Section conceded there had been numerous violations by Japanese fishermen, but that could be alleviated if the fishing zone was expanded toward the Philippines and China.[51] But China and the Philippines did not want Japanese fishing boats returning to their waters: they were developing their own fisheries.

The U.S. Fish and Wildlife Service was also uneasy about the impact of SCAP's rebuilding program. In a 1948 letter, A. W. Anderson, chief of the agency's Branch of Commercial Fisheries, asked William Herrington, SCAP's fisheries director, if "all the things SCAP wants to accomplish would be accomplished by wiping out the U.S. fishing industry, we should proceed at once to that goal? Suppose, for example, that permitting the Japanese to fish unrestrictedly and to export without limit to us would solve the problem but at the expense of reducing our fisheries to impotency."[52] If Herrington replied, the letter was not in the SCAP correspondence file.

General MacArthur had no patience with the argument that Japanese economic growth should be restricted because it might hurt the American fishing industry. It would be immoral and unjustified for the U.S. to repudiate its "international commitment that Japan would be given access to raw materials to support her internal economy," MacArthur wrote to the State Department on October 3, 1948. "Repudiation of these commitments for the fallacious reasoning given would materially weaken our moral leadership among the nations of the world. Adherence to them on the other hand would conform to traditional American policy and afford the means of advancing our positive influence in the broad field of international morality."[53]

While there may have been some token reductions of licenses in the East China Sea fleet, the overall capacity of the Japanese fisheries under SCAP had increased to an estimated 70 percent *greater* capacity than in 1939. Japan did indeed intend to use its knowledge of the sea and its fishing and whaling fleets to begin the restoration of the empire. The government was aiming to build two new fleets: one intended for the cold waters of the North Pacific, and the second to move south in search of warm water and tuna.

THE AMERICANS

An ocean away, the American fishing industry watched uneasily. SCAP was rapidly rebuilding the Japanese fishing fleet and the boats could soon be back in American waters. The Soviet Union had announced a massive five-year plan to expand fishing fivefold and to do it by 1950.[54]

The American fishing industry was weak, fragmented, and undercapitalized. Imported fish eroded prices for fish. The New England industry had struggled with cheap fish from Canada for decades; now there were increasing amounts of cheap fish from Iceland and Norway.[55] During the 1930s, canned tuna from Japan had taken over a substantial share of the American market. Congress had passed a tariff on the Japanese fish to protect the industry. But the era of tariffs to protect domestic industries from imports had ended. American fishermen found it was getting more expensive to fish in Latin American waters for tuna. Countries raised the price of bait fish and started issuing annual licenses for each crew member on an American boat. Then licenses had to be renewed after six months. The costs were increasing steadily.

From his desk at the San Francisco Academy of Sciences, Wilbert M. Chapman continued to lobby for political support to expand American fisheries deep in the Pacific. Legal scholar Harry Scheiber has documented Chapman's "prodigious campaign" to rouse public interest in Pacific fishery resources and to create political coalitions for science funding. Chapman wrote hundreds of letters and lobbied from Alaska to San Diego. He wanted to expand fisheries, but he also wanted a large-scale, government-funded, deepwater oceanography research program.[56]

"Our domestic market requires more tuna," Chapman wrote to Oregon Senator Guy Cordon.[57] What he meant was that the boats needed more water to fish. With the postwar collapse of California sardines, purse-seine boats were driven from California to Latin America, where the bait boats were already fishing. The increased costs to fish meant American boats needed to expand to new water, deep into the Pacific, where Chapman assured Condon there were "commercially abundant" supplies of fish, even in areas where the Japanese had "not yet penetrated." As was always the case with Chapman, these conclusions were presented with great confidence.

American fishermen needed new sources of fish. If they could establish a fishery in the waters around the Marshall, Caroline, and Mariana Islands before a peace treaty was signed with Japan, they could argue that Japanese fishermen should not be allowed back in the island waters.[58] It was essentially the same argument used to keep Japanese boats out of Bristol Bay salmon. "This is a matter of the greatest importance. The future of the nation's high-seas fisheries will be very largely determined by the actions of the United States Department of State."[59]

If the Americans did not act, other nations would claim "sovereign and exclusive jurisdiction" over the waters of the Western Pacific tuna fishery. Americans could be shut out of all high-seas fishing, except off their own shores. The industry needed "a source of tuna which [was] free from eco-

Fig. 3.1. Wilbert McLeod Chapman, special advisor to the State Department on Fisheries, 1949. University of Washington.

nomic or political control of adjacent countries; the only potential source was tuna in the Central Pacific."[60] Chapman repeatedly warned that Japan, which had taken 30 years to develop the fishery, was now backed by the American government and would want to go back to the fishery they had developed. If the American fishing industry was to stake a claim on the tuna resource, there was no time to lose.

The Truman Proclamation on fisheries, announced in September of 1945, had pleased neither the salmon nor the tuna industries. The salmon industry asserted the proclamation had been designed with "exhaustive consultations with and advice from the fisheries industry," but the final document had been weakened.[61] It may have been too weak for the salmon industry, but it was decidedly too strong for the tuna industry, as Mexico, Peru, and Costa Rica had immediately filed claims for expanded territorial waters. They used the American argument, creating conservation zones for their fish. For the next decade, the State Department would argue, in vain, that the U.S. had never *acted* to create a conservation zone and restrict fishing; it had just claimed that it could if it wanted or needed to. Foreign nations "willfully misunderstood" the proclamation's intent.

Any solution to this snarl of problems lay in the State Department, which, unfortunately, was preoccupied with far more important issues than fish. Under a 1944 reorganization, fishing was lumped with other "commodities" and placed within the Commerce Division.[62] The industry argued that fishery problems involved diplomacy and international law and the State Department was not watching out for the American fishing industry.[63] As the territorial claims from Latin American flowed in, the State Department was forced to confront the political problems the tuna industry had caused. By paying fees to fish, the fleet had inadvertently recognized the right of countries to regulate their waters.

The proclamation had been issued in September of 1945. Three months later, Seattle attorney Edward Allen (long involved with the salmon industry) and Miller Freeman, publisher of *Pacific Fisherman*, called an industry-wide meeting in Los Angeles to consider what the proclamation meant for the fishing industry and determine what they should do about it. It was urgent that the industry be better represented at the State Department. Inattention could cause irreparable harm. The group created the Pacific Fisheries Congress to lobby for a fisheries representative in the State Department. Through 1946, they prodded their state and federal representatives to pressure the State Department into responding to the industry's problems.

From his office at the San Francisco Academy of Sciences, Wilbert Chapman threw himself into the campaign, making the argument for industry expansion in the strongest of terms. He explicitly linked fish to national defense, to have "an American industry strongly developed across the tropical Pacific."[64] The Soviets and the Japanese understood how important their fisheries were to their national welfare.[65] The Americans did not, although Chapman lost no opportunity to explain it to them.

The United States was committed to spending public funds in the central Pacific "to gain peace and to promote and protect the welfare of the natives." The only resource the islands had was fish. An American tuna fishery in the islands would help pay the administrative costs for the territories. Failure to establish an American fishery would be a blow to overall American policy on the management of fisheries in international waters.[66] The future of the American tuna industry was at stake. If the fleet did not expand quickly into the Pacific, it would be too late. The Japanese—or the Soviets—would beat them to it. And then the Truman Proclamation could be used to try to ban American fishermen!

Chapman saw the Truman Proclamation as an interim step toward a wider policy that would embrace both the salmon and tuna issues, as well solve the problems of cheap imported fish in New England. In May of 1947, he wrote that it was necessary to recognize that many American fisheries were at their maximum point of development, especially the groundfish fisheries of New England. "The handwriting is on the wall so plain that a blind man could see it with the blunt end of his stick—and now some of the New England industry men are even beginning to see it."[67] By acknowledging that American fisheries were "mature," Chapman was laying the intellectual groundwork to justify expanding in new waters—such as the Marshall, Mariana, and Caroline Islands.

As always in these arguments, Chapman made the case that fisheries management required the benefit of science, and that governments had to fund the biological research that was necessary for management. He was a key figure in formalizing the study of the California current, started in 1948 in the wake of the decline of the sardine fishery, which grew to be the California Cooperative Oceanic Fisheries Investigations (CalCOFI), a partnership with the State of California, U.S. FWS, and Scripps Institute of Oceanography. It continues to this day.

The passage of the Farrington Act created the Pacific Oceanic Fishery Investigation (POFI) based in Honolulu, to sponsor oceanographic, biological, and technological studies of tuna, with support from FWS, the navy, and West Coast fishing interests.[68] O. E. Sette was appointed to head a South Pacific investigation, to collect data on temperature distribution in the waters where tuna could be found.[69] The $520 million Tydings Bill, an economic assistance bill for the Philippines, also passed. It included a fishery component and a five-year study of the South Pacific fisheries, with three new research vessels to do the work. The new research programs were to do the research that would lay the foundation for Americans to begin their conquest of the Pacific fisheries. It may have been a substantial commitment, but the Pacific is a very large ocean.

If American fishermen could quickly locate tuna in the Western Pacific, then under the terms of the Truman Proclamation, it would be possible to exclude the boats of other nations from the fishery, according to Scheiber. He further argues this is why Congress created POFI in 1949, and why the U.S. Navy supported fishery research bills. POFI was directly concerned with "retard[ing] active exploitation by the Japanese of the tuna fisheries of the eastern Pacific, at least until American fisheries began to make economic use of these resources."[70]

But the plan to stake an American beachhead in the former Japanese waters was not going smoothly. The flagship project, the 423-foot *Pacific Explorer*, took longer than expected to renovate the World War I ship. It also cost more. Plans had initially called for building five trawlers to fish for the ship, but that was cut to four. Finally, in January of 1947, after a final outfitting in Astoria, the *Pacific Explorer*—accompanied by 12 trawlers rigged for purse-seine fishing—set off on its shakedown cruise to Costa Rica. It was far too late in the year for the ship to set out for Alaska. It obviously wasn't ready to head west to the Marianas.

Costa Rica was already a source of tension over U.S. boats fishing off Latin America. Boats from San Diego had pioneered the fishery. They needed bait to attract tuna to their hooks, but bait was unfortunately found in inshore waters. As the catch of California sardines wildly fluctuated during the 1930s, more purse-seine boats also moved south, in competition with the bait boats. The only thing the two groups could agree on was that the *Pacific Explorer* and its fleet of a dozen modern steel boats should not be allowed to catch or even buy tuna.

The American Tuna Association, which represented the bait boats, put pressure on its congressional delegation, calling for an investigation into the contract between the RFC and Bez. Representative Thor Tollefson, chairman of the House Merchant Marine—Fisheries Subcommittee, called on the RFC to explain "why a research fishing vessel, equipped and paid for by the government, is hauling tuna out of Costa Rica waters in competition with San Diego and San Pedro boats."[71]

In June of 1947, the federal agency ordered the boat back to Astoria. The trip had lasted six months, and the ship was "lightly loaded" with 2,272 tons of frozen yellowfin tuna.[72] The vessel would sit for seven months while the RFC waited for the heat to die down. In the meantime, the corporation rewrote the agreement: the *Pacific Explorer* would go to Alaska, as intended, but it would also make a trip to the Marshall, Mariana, and Caroline Islands.

Bez stoutly defended the success of the trip to Costa Rica. It had opened up fishing opportunities for small boats to deliver tuna to motherships, while the scientific information and commercial data was made available to the entire American fishing industry. The *Pacific Explorer* could make a similar contribution in the Trust Islands. The Japanese fished with a crew between 500 and 600 people, but the *Explorer* would do the job with 250 men "living under American conditions and paid according to American standards." Bez went on to say there was "considerable confidence" that an "entire new industry will be opened up to American fishermen."[73] There needed to be no discussion of "the tremendous international importance of having these fisheries explored and as rapidly as possible developed by American nations, in the light of the Presidential Proclamation of September 28, 1945."

The voyage to Costa Rica had been a political disaster, but it initiated the *Pacific Explorer* phase in the development of American fishing power, the use of motherships to bring fish back to port for processing. "Wanted at once!" ran the ad in *Pacific Fisherman* in September of 1947. "Live bait clippers and boats to fish for tuna motherships."[74] The big ad said another mothership, the *Reina Del Mar*, was being rushed to completion and would need 40 to 50 smaller boats to maintain a rotating service between Panama, Costa Rica, and San Diego.

The poor record of the *Explorer* did not stop others from imitating the basic plan, thanks to the ex-military vessels that were sold into the fishing industry, especially in the North Pacific. The vessels were converted into what were called freezerships. Their use expanded steadily. *Pacific Fisherman* published many photographs of them, the *Neva*, *Reefer King*, and *Pacific Queen*, all ex-military vessels, all from 150 to 300 feet, and all of them wanting to buy salmon and tuna and transport it back to the U.S. for canning.[75]

The 1952 *Pacific Fisherman* yearbook proclaimed freezerships the number one technological development in the Pacific fisheries. It also pointed out the Achilles heel of the big military transport ships: the owners had intended to pack salmon in the summer in Alaska, then head to Latin American to pack tuna. But the price of tuna had dropped, as Japanese tuna flooded into American markets. "Here was a setback in plans," wrote the paper, "but one stemming from the economic situation in tuna, and not traceable at all to technological or operating conditions in the salmon enterprise."[76]

The *Pacific Explorer* wasn't the only boat Nick Bez was refurbishing; there were two ex-military vessels he renamed the *Tinian* and the *Saipan*, and an ex-navy tug renamed the *Toni B*. The ex-military vessels cost $150,000 each from the War Assets Administration and he spent another $300,000 each to refurbish them. "Bidding boldly for a place in the Light-Meat Tuna business, Columbia River Packers Association, Inc., this spring is operating two refrigerated tuna carriers in tropical waters wherever tuna are to be taken in Central and South American regions," *Pacific Fisherman* intoned in March of 1948.[77] Both were 328 feet long and would capitalize on what had been learned in the trip to Costa Rica the previous year. "*Saipan* and *Tinian* have no frills, no experiments, and would brine freeze the tuna." They would receive fish from at least six boats. Each ship carried a crew of 28 and each could make seven trips a year to and from the tuna grounds.

In August of 1947, *Pacific Fisherman* publisher Miller Freeman published a special issue, "Who Will Harvest the Pacific?" It pointed out that the Americans had to stake a claim on high-seas tuna, before the Soviets did—and the Soviets were in possession of some 26 vessels, including floating canneries, that had been supplied by the Americans.[78] Washington Representative Thor Tollefson wrote to Secretary of State George Marshall, asking what had become of the fishing boats the U.S. had shipped to the Soviet Union after the war. According to Tollefson, Lend-Lease materials sent to the Soviets included 10 cannery vessels and eight refrigerator vessels, refurbished at a cost of $21 million.

"Is the Russian fleet active in the Bering Sea area, which has been traditionally American fishing ground?" the Congressman asked. "Is it active in those waters of the Pacific to which our own industry rightfully look for a share in developing?"[79] *Pacific Fisherman* published a story the next month, saying the State Department had no idea what had happened to the Lend-Lease boats, and that it did not know if they would be used in competition with American fishermen.[80] Washington Senator Warren Magnuson followed up in April, asking why no money had been spent to build ships for American fishermen. Magnuson warned "my best information now is that Russia is about to invade American North Pacific fishing grounds with the ships which we gave them and catch fish to sell in the American market."[81] Magnuson was right, but it would take a decade for the Soviets to build the boats and launch them in Pacific waters.

When Marshall took over as Secretary of State, he realized that the issues around fisheries had the potential to derail a peace treaty with Japan, a treaty that would have to be ratified by the Senate. He authorized the position of a Special Assistant for Fisheries in April. The department wanted an attorney or someone with a legal background. The industry wanted Wilbert McLeod Chapman. Attention to fisheries policy was sporadic and ad hoc within the State Department, according to legal historian Ann Hollick. The seizing of American tuna boats could not go on. The special assistant post was established to help define U.S. policy for the benefit of U.S. and foreign officials and "to tidy up the U.S. position on foreign claims to offshore areas."[82] However, it did not have the resources to negotiate with all the countries issuing claims against American fishermen.

Chapman was awarded a Guggenheim Fellowship in April of 1948, allowing him to spend the summer examining fish collections in Europe. The School of Fisheries at the University of Washington offered him the directorship. Before the first school year was finished, the State Department made its offer. Chapman was enormously excited by the challenge. He traveled

Washington, D.C., via San Diego to get the most recent information on international fisheries problems by talking with his new friends in the tuna industry.[83]

Chapman is a controversial figure in American fisheries science. He is scorned for not being a good scientist and he is reviled for what he later became, a lobbyist for Ralston Purina Company. Both Harry Scheiber and Arthur C. McEvoy have written admiringly of his influences in the development of oceanography. He was also extremely successful during his time in Washington. He negotiated three fisheries treaties and established two international commissions, still functioning today.[84] He was deeply involved in the negotiations that led to the peace treaty with Japan, finally signed in 1951. He decisively shaped fisheries science by establishing maximum sustained yield (MSY) as American fishery policy.[85]

Chapman arrived in Washington during the summer of 1948 and found a full slate of fisheries problems. A treaty was being negotiated between the U.S. and Canada and 11 European countries. Mexico had already claimed an expanded territorial zone to protest American tuna boats fishing in its waters. There was deep unease at the rapid expansion of Soviet fisheries. And, of course, there was the biggest problem of all: where would Japanese boats fish when a peace treaty was finally signed? There was no agreement within the State Department and that meant that negotiations over a treaty were stalled. SCAP was pressing for a decision and poised to start exporting Japanese-canned tuna to the U.S.

Chapman had spent three years thinking about how to carry out the policies in the Truman Proclamation and advance American fisheries into the Western Pacific. "It is no secret that a powerful motivating force behind our Proclamation was the desire on the part of the Bristol Bay people to keep Japanese or Russian (or any fishermen) from taking fish out of a stock which we were already exploiting to its maximum safe level (or beyond)," he wrote in January of 1949 to San Francisco attorney Gene Bennett. The American home waters were depleted of fish, including the salmon, herring, and halibut stocks in Alaska. It was imperative to find new waters to expand.

Within months of coming to Washington, Chapman announced the U.S. Policy on High Seas Fishing at a Chamber of Commerce meeting in San Francisco, on December 2, 1948. It was published the next month in the State Department Bulletin. The high-seas policy was geared to address the confusion caused by the Truman Proclamation in 1945, and the string of exclusive claims it had drawn from Latin American countries. "The net result of the Proclamation so far has been to stir up a mare's nest of problems which had been quiescent and which we are barely keeping under control now," Chapman wrote in 1949. "Most of these problems are embarrassing, if not damaging, to us when the total national interest is taken into account."[86] But the policy was also written to address efforts by the American fishing industry to expand into the waters of the Mandated Islands.

"The policy of the United States Government regarding fisheries in the high seas is to make possible the maximum production from the sea on a sustained basis, year after year," the policy begins.[87] Chapman's document, published in the *Department of State Bulletin*, not a peer-reviewed journal, established MSY as American policy for fisheries management (it had also been adopted as the goal for Americans in terms of whaling).

Chapman thought that all thc nations of the world could agree that the high-seas fisheries should be kept in a state where "a maximum crop can be harvested year after year." But he acknowledged it would be impossible to agree on how the crop should be shared. "That part of the problem must be left, for the present, to free enterprise and competition. There is a crop to be taken in the international common. Each takes according to his ability. When the safe crop is taken, all stop the harvest."[88] *Pacific Fisherman* summarized Chapman's policy with a six-word headline: "No Sovereignty—Free Enterprise—Sustained Yield." The approving article declared that the U.S. would not support policies that established ownership of the seas "by any nation—itself included."[89]

Chapman believed that American fishermen were just as good as the Japanese, and better than the Soviets. There was wild optimism about expanding tuna fishing into the Pacific, and certainty that technical problems could be swiftly solved with Yankee ingenuity. But the early attempts at expansion were not going well.

A new fisheries publication, *Tuna Fisherman Magazine*, based in San Diego, began publishing in December of 1947 (the first cover had a young woman in a two-piece bathing suit pretending to be a mermaid). The magazine was an early champion of Chapman and it closely followed the industry's efforts to place him in Washington, D.C. It regularly extolled the benefits that would come from an American fishery deep in the Pacific that would generate "$80,000,000 to $100,000,000 worth of tuna each year."[90]

An editorial in the first issue was based on Chapman's writings, urging Congress to pass the Farrington Act. Time was of the essence. Under the Truman Proclamation, offshore fishing rights would belong to the nation which actually explores and exploits a fishery. "Other nations have cast a covetous eye on these waters, and it behooves us to 'git thar fustest' if we are not to lose priority in areas legitimately our own."[91] If the Americans did not act, the Soviet Union would. But it was not easy to mount an exploratory fishing operation.

The flagship *Pacific Explorer* spent seven months at the dock in Astoria before finally heading to Alaska in March of 1948 for a four-month cruise. It returned with more than $1 million in canned crab and fish fillets.[92] The voyage had been plagued by fog and difficulties in finding crab. The ship carried 200 cannery hands, paid American union wages. "With the catch too light but the wages continuing, there was no hope for Experiment No. 2 paying out," wrote the *Portland Oregonian*. "It didn't."[93]

The Reconstruction Finance Corporation issued a report in 1949 saying that $4.1 million had been spent on facilities, while the operating losses would reach $1.3 million. The RFC auctioned off the *Pacific Explorer* on November 15, 1951. The top bid, from a Portland salvage company, was for $181,387. The *Pacific Explorer* was declared a success and the U.S. Fish and Wildlife Service published several reports on the catching and processing of crab and bottomfish. The boat was sold for scrap.

"The race by U.S. firms to exploit the wealthy Pacific territories captured from Japan is underway," the *Wall Street Journal* reported on May 4, 1948.[94] Castle & Cooke of Honolulu had purchased 41 percent of Hawaiian Tuna Packers, Ltd., Hawaii's only fish canning company. They wanted to establish a fishery in the Palau Islands and one in Truk, both in the Caroline Islands. The Japanese had harvested as much as 460 million pounds of tuna from the islands during the 1930s.

The 100-foot tuna clipper *Calistar*, financed by Terminal Island Sea Foods, Ltd., made a voyage to the Palmyra Islands, and several of the Gilbert and Ellice groups and came home with a disappointing 65 tons of yellowfin and one ton of albacore. There had been more tuna, but the problem was lack of bait "and new sources of raw supply must be found to assure continued development of the $100,000,000 industry."[95]

There was certainly tuna in the waters around American Samoa and the Trust Islands as well. But getting tuna out of the water cheaply was not easy, as the famed aviator, navigator, explorer, and writer Harold Gatty learned when he tried to set up a tuna cannery in Fiji. In December of 1948, aided by a grant from the Rockefeller Foundation, he opened a tuna cannery in Samoa. The idea was that boats would be based in the Fiji Islands and deliver quick-frozen tuna to Samoa for processing. The yearly capacity was estimated at 350,000 cases of tuna "as the cannery swings into full production."[96]

"Everyone has their eye cocked toward Gatty's Samoa-Fiji deal to see how that works out," Chapman wrote in a February 1949 letter. The navy had published its regulations permitting entry into the Trust Territories and

Chapman thought Hawaiian Tuna Packers would be the only company interested.[97] The islands were too far away for any mainland tuna company to consider working there.

Developing an American fishery in these distant Pacific waters would require island bases, such as the bases the Americans were planning for the new territories they would control. Chapman saw the California tuna fleet expanding to the Marquesas and French Oceania, while the Hawaiian fleet would expand through the Marshall, Mariana, and Caroline Islands. It might take 25 years for the vision to be fully realized, but "the ground work must be laid now, while our bargaining power in world politics is at its apex. Foresight now will secure to us a rich resource in the future."[98]

CHAPTER FOUR

Tariffs

> We have been finding recently in the course of our studies of the California sardine problem that there are far more fish in the sea than had ever been suspected before.
> —Roger Revelle, 1949[1]

When a nation is wealthy, it can easily sacrifice economic gains at home to achieve political and diplomatic objectives in other countries. After World War II, the United States, so enormously dominant in so many areas, sacrificed many of its smaller domestic industries for the goal of an open market for all countries. The Bretton Woods international monetary agreement, the forerunner of the General Agreement on Tariffs and Trade (GATT), was signed in 1947 by 23 countries.[2] The U.S. was most intent that its former enemies, Japan and Germany, be reintegrated into the world economy. Agreement would be expensive.

Stability in Asia depended on a strong Japanese economy. One of the strongest legs of the Japanese economy was the tuna fishery and the apparently unlimited American appetite for tuna. The vehicle for Japanese tuna to enter the United States was the 1943 trade agreement with Iceland, which created the category of fish canned in anything except oil. "The commodity tuna canned in brine did not exist in a commercial sense prior to 1950 and it seems patent to us it was originated for the deliberate purpose of evading the 45 percent tariff on canned tuna in oil," according to a 1952 White House memorandum.[3] The trade agreements were the beginning of the institutional framework that would allow for the creation of the global tuna industry, an industry that Japan and the U.S. would dominate. Americans at the time had little understanding of how the growing global economic interdependence would affect their jobs and standard of living.[4] The process, once put into place, continues to unfold for millions of American workers today.

The battle over tuna is an interesting lens to examine the place of fishing within the economy of the two nations, as historian Sayuri Shimizu has done.[5] Both were building a transpacific trading system. Fishing was a labor-intensive, extractive industry in both countries. But fishing was a far more important industry in Japan than it was in the U.S. The Japanese fishing industry and its government were united. The Americans were not. As Wilbert Chapman realized, there was "a solid core of opposition" within the State Department and the Bureau of the Budget that kept fisheries from expanding. During the 1950s, the U.S. wanted to build up "the dollar earning capabilities of such allies as Canada, Iceland, Norway, England and Denmark, and it could do so most swiftly by giving them funds to modernize their industries and throw open the lucrative American market to the product."[6]

Starting in 1947 and over the next quarter century, the U.S. would reduce tariffs to 8.5 percent ad valorem on dutiable goods, from an average of 32.2 percent. Tariffs no longer sheltered high-wage American jobs from low-paid labor abroad. During the Cold War period, the U.S. traded export opportunities for "votes in the United Nations or goodwill in diplomatic negotiations," argued Alfred E. Eckes, a former commissioner and chairman of the U.S. International Trade Commission during 1981–90. The government sacrificed hundreds of thousands of domestic jobs to create employment elsewhere in the noncommunist world.[7]

One of the first groups of domestic workers to feel the pinch of imports from low-wage countries were fishermen and fish plant workers in New England and Southern California. The industries sought relief from Congress, but the State Department, concerned about rebuilding the Japanese economy and maintaining the air base in Iceland, brushed aside their complaints. Placing tariffs on groundfish would offend Canada, a major trading partner that had begun to expand its Atlantic fisheries to supply more fish for the American market. If the Americans put a tariff on tuna, Peru would ban fishing in their waters, potentially jeopardizing another cornerstone of postwar American foreign policy: open seas and open skies, for American naval vessels, submarines, and fishing boats.

The fishing industry was being restructured, moving from salting and drying to freezing. Decisions made a world away would increasingly impact local fishermen. The relatively straightforward relationship among fishermen, boat owners, and canners was going to become immensely complicated by geopolitical policies where fish—and fishermen—were collateral damage.[8] There was far more at stake than just fish.

For decades, fishermen in New England faced markets that were eroded by imported fish, mainly from Canada. Some 50 million pounds of fillets were imported in 1946, with the threat of more to come. The Canadian industry received government subsidies for boats, as well as processing and distribution equipment, and could produce enormous quantities of fish at prices that "completely demoralize[d] the North Atlantic fishing industry and drastically affect[ed] the fishing industry in other sectors of the country," according to a federal government survey in 1946.[9] Adding insult to the injury was a new source of cheap fish, from Iceland and Norway.

The tariff on whole imported fish, established in a 1939 trade agreement with Canada, ranged from 1.875 to 2.5 cents a pound, on a quota of 15 million pounds or 15 percent of the annual U.S. consumption in the three preceding years, whichever had been greater. For imports over that quota the agreements set the duty at 2.5 cents per pound. Fillets made up about 40 percent of the weight of a fish. By exporting just fillets, the Canadians evaded the tariff. Between 1939 and 1948, the price of fillets nearly tripled. And there was no tariff on frozen fish, which hadn't existed as a commodity when trade agreements were negotiated in the 1930s.

By February of 1951, the *Commercial Fisheries Review* reported that groundfish imports were 66.6 million pounds, the highest on record, an increase of 39 percent from 1949. Three quarters of the fish came from Canada, followed by Iceland with 19 percent, and Norway at 3 percent.[10] But the volume of fish imported into New England was dwarfed by the quantity of tuna coming from Japan.

At the end of 1948, SCAP opened a foreign trade office in New York and began to find markets for Japanese imports. "Jap crab is here again," *Pacific Fisherman* reported, using the language of the day, and adding that Geisha brand canned crab was being shipped to market through Seattle and San Francisco. "It is apparent that American high-seas policy is designed to encourage this phase of Japanese industry, even though the trade will be at expense of an American industry."[11] The Japanese had access to large tuna schools, observed the *Pan-American Fisherman*, providing them "with plenty of ammunition with which to gun down American prices."[12] The Japanese would reenter the Australian market in 1955, undercutting the developing Australian tuna fishery and prompting the industry to seek protection under a tariff.[13]

The American tuna industry asked the U.S. Tariff Commission to increase the tariff on tuna back to 45 percent ad valorum. Tuna imports were still small overall, but the industry was well aware that they would soon increase without a protective tariff, which the State Department flatly rejected.

There was an investigation and the Japanese fishing companies promptly slowed the rate of exports, much to the frustration of the tuna industry, since imports increased again as soon as nobody was looking.

The restoration of the Japanese fishing industry was costing American fishermen their jobs and risking an industry worth more than $50 million, Harold F. Cary, manager of the American Tuna Association (ATA), told Congress in late 1948. Japanese labor rates, working conditions, and standard of living made it easy for them to produce cheaper tuna, while costs were increasing for American boats. During the 1930s, the bait boats could produce tuna for $500 a ton. Now, with larger boats spending more time at sea, a ton of tuna cost between $1,500 and $2,000 to produce. The industry had grown to include 2,200 fishermen, 400 engineers, 4,000 auxiliary employees, cannery workers, and shipyard workers, and many of the jobs were unionized.[14] Crew shares for a season of fishing could amount to $9,000 a year. Skippers and engineers earned a good deal more, at a time when the Governor of California earned just $25,000.[15]

The California industry wanted Congress to place all categories of tuna under the same tariff, but Congress was unwilling to help. Under the 1943 agreement with Iceland, Iceland would have to agree to the withdrawal of an item, fish canned in anything except oil. If the U.S. acted unilaterally, Iceland could terminate the agreement.[16] The State Department firmly axed anything that could upset the government of Iceland, giving them a chance to renegotiate the Keflavik agreement. The fishing industry turned to its last resort, the U.S. Tariff Commission, asking for an increased duty on foreign fish.

Imports of tuna in brine increased 2,110 percent between 1950 and 1951, undercutting all other canned tuna on the American grocery store shelves. Even more ominous were the imports of frozen tuna, which had not really existed when the tariff rates were established in 1934. In 1946, Japan exported 4.1 million pounds of frozen tuna into the U.S. The amount doubled in 1947, and doubled again in 1948, and by 1950, frozen tuna imports were at 56.7 million pounds.[17] At least 30 percent of the American market was being supplied by Japanese tuna, second only to silk in terms of the value of Japanese exports to the U.S.[18]

Japanese tuna was so cheap that Nick Bez began sending the *Tinian* and the *Saipan* to Japan to buy tuna for $20 to $30 a ton and bring it back to Astoria to can. His total costs were $70 a ton. Buying from American boats cost $100 a ton.[19] Housewives could find Japanese-canned tuna on American grocery store shelves in 1950 for 29 cents a can. Brined tuna was even cheaper, from 19 to 21 cents a can. American-caught tuna, canned in oil, sold for between 35 to 39 cents a can.[20] To get tariff relief, the industry would have to prove that it would suffer irreparable damage from the foreign imports of tuna.

The volume of imports had increased so rapidly that the Southern California fleet had little time to respond. The price of albacore dropped to $300 a ton, from $350. Boats were tied up.[21] There were 3 million cases of tuna unsold going into 1951, depressing prices for the year and cutting down the amounts buyers would take from their boats.[22] And it was not just tuna from Japan. Mexico, Chile, Peru, Africa, Portugal, and others were increasing their exports to the U.S. Imports of tuna and tuna-like fishes amounted to 182,000 cases during the first quarter of 1949, with the majority being bonito from Peru.[23]

Opening a trade office in New York was not the only alarming thing that the Supreme Commander Allied Powers was doing. In late 1949, SCAP asked the State Department for permission to begin discussions with the Japanese about a fisheries treaty. The State Department refused, because it did not yet have a unified policy on what to do about Japanese fishing. There was a faction within the department, the internationalists, who believed that the Potsdam Conference had guaranteed Japan access to raw materials it had utilized before the war, and if that meant Bristol Bay salmon, so be it. But as state's new fisheries advisor, Wilbert Chapman repeatedly warned that unless Alaskan salmon were off-limits to the Japanese, the West Coast and Alaska fishing industry would make sure the peace treaty would have a tough time getting through the Senate. The salmon industry would make common cause with the California tuna fishermen, who were lobbying Congress for tariff protection from Japanese tuna.

Both countries were eager for the occupation to end. By 1949, the U.S. Army was still financing food relief shipments of between $350 million and $400 million a year for Japan.[24] The occupation cost the Japanese government some $900 million a year, fully a quarter of the budget. The economy was heavily dependent on exports, but raw materials were scarce and manufacturing slow to recover. Former enemies, despite U.S. efforts, refused to grant formal most-favored-nation treatment to Japan.[25] The profits from textile exports were not enough to cover food, fuel, and industrial materials, driving up costs and hindering recovery. Both sides wanted a treaty, but there were multiple obstacles.

President Truman appointed John Foster Dulles (1888–1959) as assistant to Secretary of State Dean Acheson, with responsibility to negotiate a Japanese peace treaty, in April of 1951. Two months later, the North Korean Army crossed the 38th parallel and the Korean War began.[26] The strategic importance of Japan was more vital than ever, and the U.S. government accelerated its involvement in expanding the Japanese economy. Chapman was heavily involved in negotiations around the 1951 treaty. He did his best to keep the Japanese fishing industry out of American waters, which wasn't easy with the tuna industry ready to pounce on any concession made to the salmon industry—and vice versa.

The overall goals of the peace treaty were that the Japanese should be peaceful, respect fundamental human rights, be part of the free world, and be friendly with the U.S. Japan should develop self-respect by avoiding dependence on outside charity, and it should demonstrate to the rest of Asia the benefits of the free world, helping to resist communism. Its proximity to the Soviet Union and Communist China made Japan vulnerable to communist intrigue and subversion. The policy recognized that there were natural and historic economic ties with the countries in the region, such as the U.S.S.R. and Communist China. A key concern was that Japan be able to find the raw materials it needed, without having to trade with the adjacent communist countries.[27]

The start of the Korean War changed the dynamics of the Japanese-American relationship. Prime Minister Shigeru Yoshida believed that Japan's strategic importance to the Americans had risen quickly after the North Korean attack on the South, and he tried to use Japan's elevated status to strengthen his negotiating position.[28] With the start of the war, the U.S. began to place extensive orders with Japanese companies, for ammunition, trucks, uniforms, communications equipment, and other products needed for the war effort. The U.S. also began buying fertilizers and consumer goods for other Southeast Asian noncommunist countries as part of the American foreign aid effort. During the four years that followed the outbreak of the war, the U.S. spent some $2.37 billion, much of it in Japan.[29]

Tired of the indecision within the State Department over fishing, SCAP moved in early 1950 to open discussions with the Japanese over the fishing issue. The Ministry of Foreign Affairs sent its official position on the high seas to SCAP on March 20, 1950. The statement opened with a declaration of how important the high-seas fisheries were to the future of Japan. Fishing was an essential basic industry. The statement laid out six general principles for a fisheries treaty: (1) that Japan would accept a multilateral agreement with Canada and the U.S.; (2) that Japan was ready to seek, with the U.S., a formula that would not set prejudicial precedents on other Japanese high-seas fisheries; (3) Japan would continue to honor its 1938 agreement not to fish in Bristol Bay; (4) Japan wanted full access to fisheries in the former Mandated Islands and other former possessions; (5) Japan would support the International Whaling Commission and join other international agreements for halibut and tuna research; and (6) Japan proposed that any agreement run for five years, after which any contracting party could terminate it.[30]

The demands reflected the importance of fishing to the entire Japanese economy. On one hand, Japan was dealing from a position of defeat. But the Korean War highlighted Japan's strategic importance to the U.S. Fishing was an obvious way to help restore the Japanese economy. SCAP pushed for the Japanese to return to the waters of the Trust Territories, to catch tuna for the American market. The initial two mothership operations had not been financially successful, but as long as the American tuna market was open to Japanese products, it didn't matter. There was sufficient cheap Japanese labor to ensure that tuna would undercut the American product, at least until the Japanese could improve the efficiency of the mothership operations and make them profitable.

Dulles made his first negotiating trip to Japan in January of 1951.[31] His paramount objective was to align Japan with the United States against the Soviet Union and he did not want a fish fight to delay an overall peace treaty.[32] He asked Chapman to work on a letter that Dulles could present to Prime Minister Yoshida, as the basis for an agreement on fisheries. Chapman went back to the 1938 language in which Japan had agreed to keep its fisheries out of Bristol Bay and fashioned a document that Dulles took with him to Tokyo.

Chapman's draft became the basis for an exchange of letters between Yoshida and Dulles, agreeing to remove fisheries from the main treaty negotiations. In a letter dated February 7, 1951, Dulles wrote that the Americans recognized the Japanese depended on fish for their food supply, and they thus had a special interest in the "conservation and development of fisheries."[33] After full sovereignty was restored, Japan would enter into negotiations to establish fishery treaties. This voluntary act did not imply a waiver of international rights or prohibit Japanese fishermen from fisheries in which they were currently engaged.

In return, Yoshida wrote that Japan recognized that high-seas fisheries could exhaust resources unless there was concerted action for conservation and development of fisheries. The government was aware that certain countries had adopted "international agreements and voluntary self denying ordinances to prevent exhaustion of high-seas fisheries which are readily accessible to fishermen of their own country and that if these conserved fisheries were to be subject to uncontrolled fishing from other countries, the result would be international friction and the exhaustion of the fisheries themselves." Japan would enter into negotiations in the future, and in the meantime, without waiving its international rights, would prohibit its fishermen from operating "in presently conserved fisheries."[34] Japan was officially recognizing the American claim that they had conserved salmon stocks.

In a March 5, 1951 letter, Chapman explained that the State Department had not reconciled its different views on what to do with the Japanese fisheries, but that Dulles had made a decision, based on language that Chapman had drafted. "What he decided was that he had to eliminate the fishery problem from the general problems of the Treaty of Peace or he was likely to get tangled up in such a fishery snarl both at home and with our Pacific allies that the conclusion of the Treaty of Peace would be delayed longer than was in the best interest of the United States."[35]

By spring, Chapman had decided to leave the State Department and go to work as Director of Research for the San Diego–based American Tuna Association. In April, William Herrington left Tokyo and returned to Washington, D.C., where he replaced Chapman in the fisheries advisor position in the State Department. Herrington gave a speech on April 9, 1951, at the National Fisheries Institute on Japanese tuna fisheries. Japan was heavily dependent on fishing, Herrington told the group, but there were between 250,000 and 500,000 too many fishermen, who so divided the catch that coastal fisheries faced collapse. The solution to at least some of these problems was to expand fishing for products that could be exported to the U.S. Herrington estimated there were 1,000 boats of more than 20 tons engaged in high-seas fishing for tuna and that it was certain the fleet would increase as fishing territory was expanded. If the U.S. tariff on tuna was increased to 22.5 percent, Herrington thought it would cause a temporary drop in exports, but that the Japanese industry would absorb the costs through lowering prices to fishermen and increasing cannery efficiency. "Before the end of 1951, it is likely the volume of Japanese exports will return to, if not exceed, that of 1950," Herrington warned.[36] "There is every indication that the Japanese tuna pack will continue to expand and that they can meet the increased tariff rates if they have to," Herrington wrote to Chapman on August 30, 1950.[37] It was a message that dismayed the American industry.

Herrington outlined the dilemma for American tuna fishermen and processors. Without the protection offered by tariffs, "there is very little doubt that the United States tuna fishery would greatly decline, if not practically disappear." Yet tariff protection for a domestic industry flew in the face of a national policy of reducing tariffs to increase trade, "and remember there is plenty of evidence that most people believe this policy is in the overall national interest."[38] Fishing policy was increasingly on a collision course with American foreign policy concerns, using fishing to help restore Japan's balance of payments by increasing their export of canned fish to the United States, and ensuring the destruction of the Pacific tuna processing industry.[39]

During World War II, Secretary of the Interior Harold Ickes thought a little organization and oversight from the federal government would do a great deal toward making fishing more efficient and fishermen more prosperous. Fishermen needed to modernize. The government would do its part by creating a greater demand for fish. More demand would stimulate the industry, raise prices, and allow fishermen to invest in new technologies. As fish became cheaper, the industry could sell more. Increasing seafood consumption was seen as a way for the New England industry to grow its way out of its marketing problems. Unfortunately, demand wasn't at the heart of the industry's problems, it was an inability to compete in what was becoming an international, government-subsidized industry.

Other governments were lending money to fishermen to build boats, but the Americans expected fishermen to build their own boats. Most of the larger American fisheries—New England groundfish, salmon and halibut on the West Coast, salmon in Alaska, and sardines off California—did not have more fish to catch; catches were stable or declining. Fishermen weren't making enough money to invest in new technologies, or to do the exploration necessary to find new stocks of fish to catch in places like Alaska and the Marshall, Mariana, and Caroline Islands.

During the 1930s, with the decline of traditional fisheries such as haddock, New England fishermen had pushed into deeper water, finding large numbers of a medium-sized, bright red fish that fishermen called redfish and processors marketed as "ocean perch." There are three main kinds of Acadian redfish, also known as *Sebastes fasciatus*, *S. mentella*, and *S. marinus*. The fish were available in large quantities; landings grew from 300,000 pounds in 1933 to a projected catch of 139 million pounds in 1941.[40] But would people eat this new fish? The U.S. Fish and Wildlife Service enlisted its most talented writer, Rachel Carson, to write a series of papers, geared for housewives, to tout the benefits of these new kinds of fish. *Conservation Bulletin* 33, called "Food from the Sea," was published in 1943. Three other bulletins followed.

The *Sebastes* caught Carson's attention, perhaps because they are unusual—they bear live young, just a quarter of an inch long. The eggs develop and hatch within the mother, instead of the eggs floating freely in the water. Carson wrote that in summer, before the fish spawn, "It is possible to detect the presence of young in many of the fish brought to market, because their black eyes show through the body wall of the mother."[41]

Linda Lear, in her excellent biography of Carson, said of her work in the four bulletins: "Each bulletin bore Carson's trademark of meticulous research and factual information presented in an interesting and engaging manner. In her profiles of a region's fish resources, Carson always endeavored to provide a sense of the delicate ecological balance that was necessary for reproduction and survival."[42]

Research and development had been a haphazard process; now there was a new system of knowledge, aimed at solving technical problems.[43] New technologies offered astounding tools to find and catch fish. Recording echo sounders allowed fishermen to identify schools of fish below them in the water, so nets could be set more precisely. Sonar allowed them to search the waters ahead, a technique that was especially useful for pursc scincs and midwater trawling. After 1956, the hydraulic power block would revolutionize purse seining by making it easier to haul in big nets. While the new technology greatly expanded the capacity of fishermen, it was not immediately adopted everywhere.

After decades of neglect for ocean research, Congress funded a series of bills to build research vessels. The passage of the Farrington Bill in 1946 created the Pacific Oceanic Fishery Investigation (POFI), headquartered in Honolulu and with a mission of helping to establish an American fishery deep in the equatorial Pacific, where the Japanese had fished in the 1920s. They would start by collecting data on temperature distribution in the waters where tuna could be found.[44]

POFI sponsored a tuna industry conference in Los Angeles on October 7, 1949. Presentation materials pointed to the necessity of research before a fishery could be expected to be profitable. It had taken the Japanese 15 years to develop a significant tuna fishery in the Trust Islands. The Americans wanted to do it with two ships within 10 months. "Let us hope that we may find sound ways to shorten this period between exploration and exploitation."[45] Such research and exploration was only possible at government expense. Voyages to the islands had been expensive and disappointing, but the government investigation into king crab in Alaska had been successful in helping to develop an American king crab fishery.

"If we delay in establishing fishing rights in Oceania by exploration at the least, other nations may be so firmly established that we will have difficulty in obtaining access to resources needed in the future," the meeting concluded. POFI's two new research vessels, the *Henry O'Malley* and the *John R. Manning*, would be used: one rigged for live bait fishing, the other with a purse-seine net.

"The experience of the RFC vessels *Oregon* and *Alaska* last year, and of

Fig. 4.1. Charles R. (Bob) Hitz was a crewman on the R/V *John N. Cobb.*

various private ventures in the central Pacific before and since then indicate that simply going fishing without an adequate background of facts regarding the tunas and their environment in a new region, is likely to be a costly venture without much in the way of catch in return." It was a sober note. Fisheries exploration started with optimism but finding fish was not easy, even with powerful new tools.

With the *Pacific Explorer* finally scrapped, there was the question of what to do with the four steel trawlers that had been built to fish for it. There had been plans to send the *California* and *Washington* to explore for resources off the west coasts of Peru and Chile in 1948. The vessels were crewed and ready to go when the expeditions were instead cancelled. The Restoration Finance Corporation transferred the vessels to the U.S. Fish and Wildlife Service.

Bob Hitz, a retired fisheries scientist, has traced the fate of the four boats. The *Washington* sailed on the first cruise for Seattle's newly formed U.S. Exploratory Fishing and Gear Research base on August 23, 1948, to

Fig. 4.2. The 93-foot *John N. Cobb* was built in 1950 and would serve as the primary research vessel on the West Coast until it was decommissioned in 2008. It was the smallest ship in the federal research fleet and the only vessel built of wood.

the Bering Sea with the *Pacific Explorer*. The *Oregon* sailed on an albacore exploratory trip on Exploratory Cruise 3 for the Seattle base, and then she was transferred to Pascagoula, Mississippi, to the newly formed Gulf of Mexico's exploratory base. The *Alaska* was transferred to the State of California as a research vessel, and apparently the *California* was sold to a fisherman in California.[46]

The *Washington* was not well-suited to serve as a research vessel and ownership was transferred to the Republic of South Korea. The Koreans paid $100,000 for the boat and the Fish and Wildlife Service built the boat that it really wanted, the *John N. Cobb*. Named after John Nathan Cobb (1868–1930), the founder of the School of Fisheries at the University of Washington, the 93-foot wooden *Cobb* would become the heart of West Coast fisheries research for the next five decades.

"Exploratory fishing occupies a new and important place in Government aid to the fishing industry," wrote the Fish and Wildlife Service in 1949. "Our fishing industry is now beginning to receive, through these exploratory vessels and the technical personal aboard them, the assistance long rendered agriculture by extension services and regional research stations."[47] Increasing the catch from the ocean would be little different from increasing the yield in agriculture. The *Cobb*'s shakedown cruise was to Alaska to look for pink shrimp (*Pandalus jordani*).

Heading the Seattle exploratory gear base was a young scientist and war veteran, Dayton Lee Alverson, who graduated from the College of Fisheries at the University of Washington in 1949. It had not been long since that trip to sea out of Astoria on the *Harold A*, the trip that convinced him that he wanted to be a marine biologist. Together with George Yost Harry Jr. and Jergen Westrheim, Alverson was engaged in the systematic study of the amazing variety of rockfish that fishermen were finding in increasing numbers. The size of the population had to be astounding. Alverson began working with West Coast trawlers, trying to find out where they were catching large trawls of rosefish and its many cousin species.

AMERICAN SAMOA, 1945–55

The U.S. did not directly annex the Japanese Pacific islands. Instead, the entire northern Pacific Ocean area became a trusteeship of the United Nations, administered by the U.S. The area was under the supervision of the UN's Security Council, not the General Assembly. Since the U.S. possessed a veto in the Security Council, the islands were under absolute American control.[48] Through its control of the Trust Territories, the U.S. had access to isolated sites deemed suitable for nuclear testing.

The Americans saw little of value in the islands, except that their remoteness and small population made them ideal for nuclear testing. Having wrested control of the Pacific from the Japanese, the Americans intended to use the islands as a base to dominate the Pacific, projecting American power.[49]

The American Pacific Proving Grounds included Bikini and Eniwetok atolls in the Marshall Islands, Christmas Island, and the Johnston Atoll. Between 1946 and 1952, the U.S. conducted 110 atomic bomb tests. Nuclear testing had to be done in areas with few human residents and where the oceans could absorb the radioactive debris. Tests had to be kept as secret as possible. The U.S., France, and Britain concluded the safest place to accomplish these goals were small islands in the remote Pacific. The testing programs were carried out without the consent of native peoples. "The perceived strategic necessities of the Cold War power balance outweighed many environmental concerns that now seem absolutely essential."[50]

The residents of the Bikini Atoll, a group of 23 islands, were evacuated before the tests began. They were told they could eventually return home, and then moved one hundred miles to the Rongerik Atoll; they were moved again to Kwajalein, still farther away, then to the Kili Islands in the extreme southern part of the Marshalls. Bikini had an extensive lagoon for fishing and good anchorage for boats; the other sites did not. The Kili anchorage was poor, it was not protected, and landing conditions were only favorable during part of the year.[51]

The Bravo Test at Bikini Atoll in March 1954 was the most powerful bomb ever exploded. Unfavorable wind conditions and the unexpectedly high yield of the bomb caused the worst radiological disaster in U.S. history. The Marshallese people living on four atolls were blanketed with radioactive fallout, as was the crew of a Japanese fishing boat, the *Lucky Dragon*.

The Americans wanted the remote islands for nuclear testing, but they were not really interested in the islands themselves. They were officially on record as opposing the imperialism of the Great Powers. But it was more important to contain communism, and that meant that the islands, now in the possession of the United States, needed to be modernized. While other nations after World War II dismantled their colonial empires and allowed new nations to move toward self-rule, the Americans delayed that process, especially in the Pacific.[52]

The Americans portrayed themselves as not motivated by exploitation, but by the well-being of the local inhabitants. Francis B. Sayre, the U.S. representative on the United Nations Trusteeship Council, wrote frankly in 1948 that the American objective in the islands was to turn them from "international liabilities into assets."[53] If the islands possessed assets, they must be utilized for the common good of the United Nations. "The former Japanese mandated islands are of little or no economic importance," wrote Sayre, but he added that they were of enormous strategic value, which far outweighed any social or commercial considerations.[54]

The initial American policy was driven by a desire to build an economy that would support modernization of the natives.[55] One of the few potential economic assets the islands had was access to tuna. The navy had built a cannery at Pago Pago during the war. In 1952, the Department of the Interior went looking for a company to operate it. William D. Moore Jr., foreign production manager of the Van Camp Sea Food Company in San Diego, went to take a look. He was not impressed. The refrigeration plant was in poor repair and there was no fishing fleet to supply the plant. The company would have to pay workers an American pay scale, but Moore figured it was worth a gamble. He signed a five-year lease and the company set about upgrading the plant and getting ready to can tuna. Van Camp agreed to provide for the improvement of American Samoa by developing the skills of fishing and fish processing among the Samoans.[56]

Fishing is both difficult and dangerous, and so is processing. "They fish during the day, unload their catch at night, and even catch bait at night when necessary," wrote Peter T. Wilson, a cannery official, in his memoir published in 2011. "They learn to go without sleep, eat standing up, go into cold, bloody water in the bait wells late at night to get the tuna unloaded, pull anchor ropes and nets while still half asleep, hook tuna until their backs are ready to break—and then hook some more."[57]

When canneries are in production, they must function 24 hours a day. The multistep process involves difficult manual labor at every stage. Frozen fish are unloaded onto broad platforms under the blazing sun. Workers must hunch over to sort the fish by size with a hatchet-type pick. Another group of laborers transport the fish into the freezing cold storage warehouses. The fish has to be defrosted and deboned, then swiftly canned, sterilized, and packaged. The cannery worked on 12-hour shifts and the work was arduous and dangerous. There was the heat of steaming processes, the quick and sharp knives for processing, and the powerful and potentially crushing packaging machines.[58]

With modernization of the islands in mind, the U.S. Tariff Schedule in 1953 approved a clause, under headnote 3(a), that products from American Samoa could be exported to the U.S. if the local component was at least 30 percent of the value. A second powerful inducement was that the territory was exempted from the Nicolson Act, which prohibited the landing of fish by foreign flagged vessels in U.S. ports.[59] The first year, Van Camp found that Samoans could catch, clean, and process fish and seafood, "but it takes three to five people to do and perform what one American laborer or worker in the States can do."[60] If companies were to move to the island and build an economy, the federal minimum wage had to come down. Van Camp's was operating at a loss and no other industries had moved to the islands.

Van Camp appealed to Congress in 1956 for permission to lower wages, permission that was granted later that year. From San Diego, the American Tuna Association opposed the lowering of cannery wages. Instead, the association argued wages should be increased to bring them in line with American wages, lessening the already large competitive advantage the Samoan cannery had over American operations.[61] From Seattle, Nick Bez agreed, anxious to continue importing cheap Japanese tuna to the mainland. After a lengthy set of hearings, Congress agreed to lower the wages. StarKist Samoa, Inc., began operating a new cannery adjacent to the Van Camp plant in 1963. Both canneries depended entirely on foreign boats to supply the fish.[62]

Under the new tariff regulations, fish caught by non-American vessels could land their catch at Pago Pago, where it would be processed, then exported to the United States with no tariff.[63] This was a very large carrot for tuna companies, since tuna carried a 35 percent duty for an oil pack, and from 6 percent to 12.5 percent if canned in water. In January of 1954, StarKist contracted with seven Japanese tuna vessels to deliver their catch to the cannery. Just 338 metric tons were delivered the first year, but landings soon increased to 15,588 metric tons by 1965, as vessels from the Republic of Korea and the Republic of China joined the fleet.[64] They also contracted with a Japanese firm to charter the *Saipan Maru*, a 3,700-gross ton cannery vessel, to freeze and transport tuna from 20 Japanese fishing vessels and bring it to Samoa for processing. They would pay $275 a ton for albacore and $190 a ton for light-meat tuna.[65]

With this series of decisions, American Samoa became the focal point for Japanese tuna to enter the United States, ultimately leading to the demise of tuna canning in San Diego, San Pedro, and Astoria. With the establishment of a tuna cannery in Puerto Rico—also heavily subsidized and exempt from American wage standards—by 1953 the stage was set for Pacific fishermen to move into the Atlantic and Indians Oceans. The Japanese boats would successfully develop the longline techniques during the 1960s to target Atlantic bluefin tuna, (*Thunnus thynnus*), especially off Brazil. They would eventually move on to the Gulf of Mexico, the only known spawning area for bluefin in the Western Atlantic.[66]

The tariff decision benefitted the Japanese boats; it did nothing to help American boats because it was not economically feasible for them to land their catches in Samoa, according to Wilbert Chapman, now keeping an eye on the Japanese from his ATA office in San Diego.[67] He thought the Japanese motherships delivering tuna to Pago Pago for processing were probably losing money, but continued because of the dollar exchange from the tuna sales. The ATA urged the Department of the Treasury to file a complaint that Japanese tuna were being dumped in the American market, but the agency refused to do so.

The tariff decision allowed Van Camp to can frozen Japanese tuna and ship it to the U.S. at no tariff. "This decision made it possible for Van Camp to contemplate a profitable canning and frozen tuna business in Samoa using cheap Japanese fish. This they embarked on this in 1954 and have continually expanded. There is as yet no noticeable limit to the expansion" wrote Chapman.[68] By 1955, products from the Van Camp Sea Food cannery accounted for 80 percent of the island's total exports.

The ATA considered trying to protest the decision but that would require an amendment to the relationship between the U.S. and Samoa, which was acquired by the U.S. through a treaty with Germany and England that allowed the two countries to retain navigation and harbor rights. If the tuna industry couldn't get a bill out of a Senate committee, it was unlikely to get a treaty changed.

"Accordingly, the Japanese have opened up a substantial breach in our protective wall which we are not now prepared to mend." Chapman wrote. "This enables them to fish in areas of the ocean we are unable to economically work by using our ports. It enables them to get fish into this market at prices with which we cannot compete, and which are actually cheaper than fish of the same kind and quality which is imported directly from Japan to the United States."[69]

The cost of a case of yellowfin produced in Samoa was less than half of San Diego production costs. Chapman estimated that Van Camp was producing 10 percent of its total output in Samoa and "this directly reduces the firm's payroll in Southern California by about that amount." The Samoan plant handled about 6,000 tons of tuna in 1956, equal to what could be caught by six or seven tuna clippers, thus impacting the jobs of 70 to 90 fishermen.

"A third effect of this has been a solid contribution to an almost catastrophic reduction in the price per ton for tuna received by fishermen in Southern California. Before the Samoan plant started operations in 1954 the price of tuna in Southern California was $350 per ton for yellowfin tuna and $310 per ton for skipjack tuna. Today these prices are, respectively, $270 and $230 per ton."[70] Frozen tuna prices also dropped.

As cheap Japanese-canned tuna flooded the shelves of American supermarkets, American fishing boats were tied to the dock in San Pedro and San Diego. The mainland tuna processing plants began to close, and more Japanese fish was delivered to Samoa.

ICELAND

No country moved more aggressively after World War II to evict the foreign fishing boats in its waters than Iceland. It declared a 200-mile limit in 1948 (based on the Truman Proclamation, especially irritating the State Department). It moved incrementally to enforce this 200-mile boundary, claiming four miles in 1952, to 12 miles in 1958, to 50 miles in 1972, and finally to 200 miles in 1976. The first expansion, to four miles, set off an international process of meetings, culminating in the three Law of the Sea conferences that form the framework of modern fisheries management.

With the end of World War II and an agreement on when the Americans would finally leave Iceland, the Icelandic government concentrated on growing its market for its own fish. Foreign boats fishing in Iceland's waters had been an irritant since the British steam trawlers arrived in the 1880s. Now there were British, French, and even German trawlers fishing in Iceland's waters, producing fish that competed with Icelandic fish on the world market.

Iceland's delegation to the United Nations requested in 1952 that the International Law Commission study the issue of territorial claims. There was strong opposition from the Western European nations fishing in Icelandic waters, but the UN backed Iceland's motion. It was an important milestone in the evolution of the Law of the Sea, according to Icelandic scholar Hannes Jónsson. "By adopting this resolution, the General Assembly implicitly agreed with the Icelandic delegation that the opinion of the leading West European powers—that the 3-mile limit was firmly established rule of international law of the sea—was wrong, and accepted the necessity of studying the regime of territorial waters."[71]

Iceland wasn't the only country unhappy with foreign boats in their waters. Despite decades of negotiations, Norway was unable to reach agreement with British boats that had fished in their waters since 1906. Norway acted unilaterally in 1935 and issued a decree expanding its waters to a four-mile belt seaward from straight baselines drawn between 48 points on Norwegian promontories, islands, and rocks. The result was the enclosure of a large area of water that had formerly been regarded as the high seas and an important British fishing ground. Britain applied to the International Court of Justice at The Hague for a ruling, and in 1951 the court upheld the Norwegian position.[72] The decision was one of the first successful attempts by a country to restrict foreign fishing on territorial grounds, and it was a substantial blow, not only to Britain's distant-water fleet, but to the U.S. distant-water claims as well.

Iceland moved swiftly to reinforce the decision. Britain was Iceland's second-largest market, after the United States, annually receiving a quarter of Iceland's fresh fish.[73] On March 19, 1952, Iceland closed Faxa Bay, an important fishing area, for conservation reasons. Iceland's coast guard arrested a British trawler on July 19, claiming it was fishing inside the new fishing zone demarcation line.[74] Iceland's decision expressly linked conservation with enclosure and protection from the distant-water fleet. The British fishing industry responded by banning Icelandic trawlers in British harbors.

What the British industry overlooked was that the Faxa ban also applied to Icelandic trawlers, not just to the foreign boats.

The outbreak of the Korean War in June of 1950 had reverberations in Iceland, according to historian Valur Ingimundarson. The presence of Soviet fishing boats off Iceland prompted the government to ask the U.S. to beef up the island's defenses. About 3,000 American troops arrived to do air and sea patrols of the North Atlantic. The U.S. Air Force pressed hard for a second base, but Iceland was not willing to allow another American base.[75]

The Americans were displeased that Icelandic fish was being sold in the Soviet Union. Iceland firms began exporting halibut in late 1951 and small amounts of fillets were starting to enter U.S. markets. Ingimundarson cites a State Department document in which President Eisenhower suggests that the Americans should buy all of Iceland's fish production and give it to poor countries as a grand humanitarian gesture. "The suggestion was not taken up lest it should open a floodgate to similar requests from other countries."[76]

CHAPTER FIVE

Industrialization

Aside from the problems of finding consumers for "underutilized" species, it frequently turns out that in the ecological sense the species is not underutilized at all.
—Christopher Weld, 1979[1]

The fishing world changed dramatically one March morning in 1954 when the 280-foot *Fairtry* slid down the marine ways at the John Lewis and Sons shipyard in Aberdeen, Scotland. It was more like an ocean liner crossed with a whaler than it was like a fishing boat. There were photo spreads in all the fishing periodicals, as well as the British daily press, because the *Fairtry* represented something entirely new: it brought together, for the first time, modern fish catching, processing, and freezing capacity. It could stay at sea for weeks at a time and fish in gale-force winds of up to 63 miles an hour.[2] Its first trip was to the Grand Banks off Newfoundland for cod; during its first year of operation, it produced 650 metric tons of cod fillets from 2,000 metric tons of cod in just 37 days. When it got back to London, its owners had to replace the filleting equipment with machines that could handle smaller fish; by 1954, Newfoundland cod were smaller than they had been in the past.[3]

The *Fairtry* was the result of six years of experiments begun in 1947 by the British and Norwegian whaling company, Christian Salvesen. The work started with a former Algerian class mine sweeper, the *H.M.S. Felicity*. Rechristened the *Fairfree*, it was a floating laboratory for Sir Charles Dennistoun Burney, R.N., an enthusiastic yachtsman, aircraft engineer, and inventor of the paravanes used in World War II minesweeping. He was interested in a radical new design for trawl doors that he hoped might lift ground trawls free of the bottom and permit fishing at various depths of the water column.

It took more than a decade to build what came to be called the midwater trawl, one of the most significant technological advances in twentieth century fishing technology. After six years of experiments with the 200-foot ship, Salvesen planned a new version: 2,600 gross tons, with an overall length of 280 feet and accommodation for a crew of 80. The largest British side trawlers then in use averaged 185 feet and 700 gross tons, with a crew of 20.

Salvesen moved into fishing because the company foresaw the decline of whaling. They brought the stern ramp from the whaler and placed it on the *Fairtry*. It simplified the operation of the trawl and allowed the use of much larger nets than those that could be handled on conventional side trawlers. It allowed trawlers to be considerably larger, with higher sides. That allowed for the introduction of more sophisticated processing equipment. During the 1950s and 1960s, most of the vessels were around 1,000 gross register tonnage (GRT), but new ships in excess of 3,000 GRT soon followed. In order to cover costs for such expensive vessels, factory ships had to concentrate on the most productive fishing grounds and resulted in the spread of fishing throughout the oceans—not only on or near the bottom at depths of more than 1,000 meters, but also at all depths between the surface and the bottom.[4]

The first Soviet boats moved into the waters off Newfoundland in 1954; perhaps that is where they first saw the *Fairtry*. Or perhaps they had followed the stories of the ship's development in the trade press. At any rate, the Soviets placed an order with Brooke-Marine, Ltd., of Lowestoft, England, to build 20 modern fishing boats, at a cost of $20 million. The contract lent some insight into the state of Soviet fishing. The British engineers received vessel specifications that were based on designs from the 1930s. "They seemed to have no knowledge of what went into the making of a modern fishing trawler."[5] A three-volume American study of Soviet economic development, published in 1973, concluded there was no fundamental industrial innovation between 1917 and 1965. "Soviet innovations have consisted, in substance, in adopting those made first outside the U.S.S.R. or using those made by Western firms specifically for the Soviet Union and the Soviet industrial conditions and factory resource patterns."[6]

The British trawlers, with the latest technology, were turned over to shipyards in the Soviet Union, Poland, and East Germany, to be used as prototypes for a fleet of new boats. The first of two Soviet factory processing clones, the *Pushkin* and the *Sverdlovsk*, arrived off Newfoundland in 1956. Two dozen *Pushkin*-type factory ships were fishing in another two years. By 1958, new fisheries with enormous catches were being established from Maine to Greenland to the Barents Sea.

The ships were 280 feet long, could fish in almost any sea, operate continually for 10 weeks, and can, freeze, and make fish meal. They carried a crew of around 100 people.[7] The huge engines could haul enormous nets fishing at great depths, deeper than fishing boats had ever fished before. "Although knowledge of the region and of the fish biology was insufficient, the exceptional wealth of this region and the sharply delineated frontal zones (which permit accumulation of fish) facilitated large catches by the fishing fleets of several countries," Soviet scientists wrote.[8]

The 20th party congress in 1956 voted to accelerate construction of stern trawlers, with an investment of more than 10 billion rubles between 1956 and 1975, resulting in the construction of the world's largest fishing fleet. They built more than 5,400 distant-water vessels and amounting to at least half the world's gross vessel tonnage for fleets of this size and type.[9]

How much did the Soviets spend on their fleet? There are various figures in the literature. American scientist Al Pruter wrote that between 1946 and 1950, the Soviets invested 447.4 million rubles in expanding its fisheries. T. S. Sealy, a translator working in marine technology, published a paper in 1974 estimating the Soviets spent 4 billion rubles between 1946 and 1965.[10] Two American scientists argued in 1968 that approximately 3 percent of the total Soviet capital investment in industry went to the fishing industry, with three-quarters going to build catcher vessels.[11] Whatever figures were released might not be accurate, cautioned Soviet analyst Terence Armstrong. The Soviets wanted to show themselves as modern and efficient.[12] They heavily subsidized the development of a distant-water fishing fleet based on a number of factors, including the need for a protein supply that would be cheaper than agriculture, providing employment, replacing imports, and generating foreign exchange.[13]

The Soviet expansion into fishing was initially enormously successful, beyond anyone's wildest dreams, until the stocks began to falter. Fisheries not only contributed greatly to the food supply, but employed many workers involved in shipbuilding, transportation, distribution, and food production sectors of the economy. It supported a large, centralized state bureaucracy and a substantial educational establishment.[14] The focus was on the discovery of new, unexploited stocks, and the gear needed to catch and harvest them effectively, such as the deepwater fishery the Soviets pioneered in the North Atlantic for grenadier (*Coryphaenoides rupestris*), a fish distantly related to cod. The Soviets caught 15,000 metric tons in 1969 and that grew to be 82,600 metric tons two years later.

If you were going to design a fishing system that would spend a virtually unlimited amount of money to catch as many fish as possible as quickly as possible, that would be the Soviet system of the 1950s and 1960s. Fleet operations were centrally controlled and coordinated. Information about water temperature, plankton, and other indicators of fish were circulated daily around the fleet, along with catch figures. The commodore had reconnaissance ships, one for every 10 to 20 catching ships; some fleets had an aircraft. The reconnaissance units located the fish, the catcher boats were sent into the area, and the explorer ships moved on to find new sources of fish.[15]

Soviet scientists bragged that the ships shared information openly with each other; it was sign of worker solidarity. "From fear of competition, capitalists carefully guard their production secrets and agree to exchange them only for a high fee." Soviet scientific-technical collaboration was carried out under "different principles and presents a new attitude in international relations. The Soviet Union gives foreign governments unencumbered aid in the development and reconstruction of fishing industries for raising the living standards of the population."[16]

Modern technology had been introduced in fisheries elsewhere, "but the translation of the ideas into large-scale practice is a wholly Soviet achievement."[17] The planners concentrated on building and managing the fleets. They increased production through additional vessels and the application of more sophisticated technology. Another important feature was standardization of design. There were 20 kinds of boats and eight marine engines.

Soviet newspapers publicized the competitions workers held to fulfill their quotas early, or exceed them. When the fishermen in the Far East fulfilled their 1962 quota by 10,000 metric tons of fish, there was a competition to fulfill the plan for 1963. "It is agreed that the fishermen of the Far East shall fulfill the yearly plan by December 1, 1963, and catch not less than 50,000 metric tons of fish above the quotas." Each area was expected to produce more tonnage. The fleets pledged to increase output, more quickly, more efficiently, and at lower production costs. Workers ashore pledged to decrease the time for loading and transport ships by an average of 10 percent.[18]

"The biologists in the Far East promise to find concentrations of fish in new areas of the Pacific and Indian Oceans and give practical help to the industry in regard to catching and preparation of the catch of such valuable species as mackerel, Spanish mackerel, tuna, halibut, saury, and other fish."[19] The bonus could be substantial, 24 percent above the regular salary. Workers onboard the whaling and fishing fleets received their food, so salaries were particularly high, especially in the early years when catches were high. In addition to the bonuses, there was recognition and rewards.

Failure to meet the targets was punished: captains could be demoted, and workers who performed poorly were not hired for the next season. Wastage was systemic, with more whales and fish being killed than could be processed before they spoiled.

Over the years, information about the Soviet fisheries has emerged. Soviet fishing Captain Vladil Lysenko published his memoir, *Crime against the World*, in London in 1983. There are also the memoirs of Soviet whaling Captain Alfred A. Berzin. Both tell very similar stories.

Lysenko described himself as an old deepwater fisherman who had been a captain in the Soviet fleet for 17 years. He fished out of Murmansk in 1953 on a trawler that could hold a maximum load of 275 tons. Yet the production target per voyage was 300 tons. About three-quarters of the fish would be salted, passing through the hands of below-deck crew of three men, who had to crush ice and pack each fish individually. The ship worked through storms, because time spent not working was not paid, "so work only stopped when the weather got so bad that the trawls started to break."[20]

The Soviet Union flagrantly violated territorial limits of waters where it fished. Taking immature fish in their trawl nets led to the overfishing of the Barents Sea. Petty officials had the power to squeeze the crews; Lysenko described one drowned sailor's widow being charged for the clothing in which he died. He describes the corruption of a minor official who cheated the crew by issuing substandard supplies. The men used long, sharp knives but with no safety gloves. Gloves would be ordered, but were poorly made and offered little protection. Men were under constant nervous strain.

The processing plants had to buy the fish at a state-set price, so there was no attempt to maintain quality. "The entire Soviet fishing industry was based on the principle of squeezing as much work out of the wretched trawlermen as possible and paying them as little as possible."[21] He compares working on the back deck to being in a penal colony doing hard labor. The guaranteed wage was only paid for time spent at sea or in transit, not when the boat was in dry dock for repairs. Living conditions for families of sailors were poor. He describes targeting young herring off Norway, fishing illegally within three miles, where the herring schools were denser, but the fish were immature. The trawl nets could catch 40 tons of herring after 20 minutes of trawling. Fish that had not yet been processed were thrown back into the water, where they would rot on the ocean floor, poisoning the grounds.

"Thus the notion that the fish stocks were practically inexhaustible was highly convenient for those people who preferred to shut their eyes to the future," Lysenko wrote.[22] By 1955, he noticed that fish were not as plentiful. Such rapid expansion soon outgrew port facilities—trawlers had to wait to unload and the fish would spoil. "It was then that I began to realize that very little of the fish we caught actually reached people's stomachs."[23]

Soviet whaling scientist Alfred A. Berzin told a similar story, writing that Soviet whalers ignored "every quota restriction or prohibition agreed on by the International Whaling Commission (IWC), the U.S.S.R. factory ship fleets killed every whale they could find."[24] The crews were under pressure to meet annual targets and only paid a bonus if the target was exceeded. But the targets steadily increased as the whale populations shrank. Scientific reports that populations were crashing were ignored. Data was widely falsified.

"Soviet cheating started small and quickly ballooned," writes historian Kurkpatrick Dorsey.[25] In 1946–47, the Soviets used Norwegian gunners on the fleet of vessels supplying the floating factory ship, the *Slava*, but the gunners would not return because conditions were so bad: poor food, an inefficient crew, always behind schedule, and at least one Norwegian died under suspicious circumstances. The Soviets hunted out of season, took all whales that crossed their paths, including calves, and reported false data to the IWC. In addition to false numbers, they reported inaccurate ratios of males to females, average sizes, and date of capture, to cover up for hunting out of season. Of the 234,000 whales killed by the Soviets between 1948 and 1972, only 140,000 were reported. The falsification of the data made it impossible to make accurate conclusions about the size of whale populations.

Berzin wrote that in theory, catch levels were supposed to be sustainable. In reality, it was an unregulated system that consumed natural resources with considerable waste, exacerbated by the fact that output was not based on genuine needs or national demand, or constrained by mounting costs and the need to show a profit.[26]

The Soviet whale catch data was routinely falsified. It was not until 2010 that whale numbers were sufficiently analyzed to correct the Soviet data. Between 1946 and1986, the Soviets reported catching some 185,000 whales, while the actual catch was 338,000. It is also known that Japan misreported catch data from its coastal whaling stations on two species (sperm whales and Bryde's whales, *Balaenoptera edeni*) until at least 1987, but it was not on the scale as practiced by the Soviets, nor was there a planned system of deception authorized at high levels of the government.[27]

THE UNITED STATES

Fishing was being revolutionized on many fronts, and especially in the implementation of freezing technology. Through the 1930s, developments in marine refrigeration allowed boats to fish further from home and stay at sea longer. Governments facilitated the expansion of fishing by building refrigeration plants, helping to expand the new markets for frozen fish. Clarence Birdseye (1886–1956), the Gloucester, Massachusetts, businessman pioneered the introduction of "quick-frozen" foods. The development of large open-top freezers allowed bigger grocery stores to dominate by offering more frozen food. Swanson's came up with frozen individual meals, marketed as "TV dinners." Government home economists set about proving that not only did processing foods save housewives time, it saved money.[28]

There had been methods of freezing in the mid-1920s, but Birdseye's major contribution was packaging. Working with DuPont chemical company, he developed moisture-proof cellophane wrapping that allowed foods to be frozen more quickly. It was extremely useful for fish, creating a neat package that did not leak. The market for frozen fish expanded rapidly. Fish fillets could be flash-frozen in large blocks in Canada, Iceland, and Norway, and shipped to the U.S. where the blocks were sliced, breaded, deep-fried, and frozen again. The fish stick was introduced on October 2, 1953. The sticks were uniform, simple to prepare, and, best of all, required no cooking (merely heating), totally divorcing fish sticks from the idea of messy, smelly fish that consumers had trouble cooking. Imports of fish blocks soared to 50 million pounds by 1956.[29]

The fish stick is the ocean hotdog, as historian Paul Josephson put it. Three years in development, it signaled "the modern era of easy-to-prepare, nutritious foods."[30] Many innovations in technology came together to create it: the development of floating factory processing with their massive catch potential, new refrigeration techniques, and the spread of the supermarket with its frozen food aisle. Governments and industry sponsored education campaigns to get consumers to try the new products.

They were wildly successful at getting Americans to eat more fish, but that didn't help New England's fishermen. Catches had declined, they had to fish further from home and they could not compete on price with fish blocks from Canada and Iceland. The government thought the growing demand for the fish sticks would solve the problem. But the problem was not lack of markets; it was the growing globalization of fishing and the modernization of air transport. Local fish could easily be replaced by a cheaper fish caught by government-subsidized boats with a low-wage crew, operating under the protection of government policies linked to foreign policy objectives. It was not only New England's fishermen that suffered because of the tariff policy. New England lost 150,000 textile-related jobs between 1929 and 1950 as hundreds of mills closed.[31]

Fishing was being industrialized, along with the rest of the American agricultural system. Agriculture became closely woven with domestic and international policies in this postwar period. Industrial agriculture developed quickly, with federal money to build dams to provide irrigation. Ideas about agriculture were transferred to the oceans: fish were a crop that could be harvested "like radishes," and could sustain high harvests over time.[32] Trawl nets plowed the ocean floor, just like tractors plowed a field, stirring up nutrients—but also breaking fragile ocean-floor life forms, a consequence that was not revealed until the 1990s, when cameras attached to fishing gear revealed the negative impacts of trawling on the ocean floor. These technologies and ideas about nature were exported to third-world countries, which were encouraged to industrialize their fisheries.

The New England industry pleaded that it could not compete against fish from Canada and Iceland.[33] It pushed for a higher tariff, but a report by the State Department found that imports had not had a negative effect. The U.S. Tariff Commission finally held a hearing in 1951. Three of the five commissioners voted against relief, arguing that the industry showed adequate profits, rising wages, and increasing production.[34]

Imports from Japan were also threatening the Southern California tuna industry, with its union-scale processing plant jobs. The vessels themselves had large, unionized crew, and boats had to travel farther to find full loads of fish. The American Tuna Association, based in San Diego, knew it needed to make some changes if it was to survive. The group reorganized itself and offered Wilbert Chapman a job. After three years in Washington, D.C., Chapman had resigned and was going to take himself back to his job, director of the School of Fisheries at the University of Washington. When the ATA offered a job that would keep him in the heart of the political fight to expand American fishing deep into the Pacific, he took it. About 90 percent of the fleet was tied up at the dock, yet fishermen voted to tax themselves to raise $600,000 a year to promote American canned tuna.

American fishermen had to block a proposal to cut the tariff on tuna canned in oil, their premium product, from 45 percent to 22.5 percent.[35] "We are going ahead because we are being hurt too badly to stand still," wrote Chapman.[36] Fishermen staged protests with their boats, and their wives organized to support Point Loma restaurants that served American tuna. Letters poured in to Washington, D.C., urging lawmakers to increase the tariff on canned tuna.[37] Eight bills were introduced in Congress seeking to impose various tariffs on canned tuna. A subcommittee, after two days of hearings in October of 1951, recommended an emergency tariff of three cents on fresh and frozen tuna.[38] Such a small increase would be easily offset by the lower Japanese labor costs.

The Tariff Commission held a series of hearings in early 1952. The industry had to prove that imports were hurting them domestically to qualify for relief with a higher tariff. The collapse of California sardines had driven about 100 purse-seine boats into the tuna fishery. There were 200 bait boats in the ATA, director Harold Cary told Congress. Imports threatened the very life of the industry, both fishermen and processing plant workers.[39]

Not all of the tuna industry supported tariff relief. Nick Bez refused to support the bill. The Columbia River Packers Association opened a trade office in Tokyo in 1951 to buy frozen Japanese tuna, which it shipped in the *Tinian* and the *Saipan* to Astoria. There was no tariff on frozen tuna. The company argued that it was unable to buy all the tuna it needed from Oregon and Washington fishermen, and if the company was to stay in business, it had to buy fish elsewhere.[40]

The State Department also objected, concerned that a tariff, even of three cents, would seriously impair Japan's second most important export (silk was first). The Latin Americans, who were already seizing American boats found in their waters, would certainly resent it. The department suggested a compromise, allowing the first 25 million pounds of fresh and frozen tuna into the country without a tariff.[41] The Japanese government also protested the tariff, but announced that it would voluntarily place a quota on its tuna exports to the U.S., effective on May 1, 1952. Exports would be limited to 12,000 tons of raw fish and 1 million cases of canned fish. In return, the Ministry of Trade and Industry asked Congress not to vote on import duties. It was a clever maneuver, postponing any finding that Japanese imports had hurt American markets.

Congress placed an emergency three-cent per pound tariff on all imports of fresh and frozen tuna, a temporary measure that would last for a year while the Secretary of the Interior and the U.S. Tariff Commission investigated the situation, with a report due to Congress by January 1, 1953. But the bill was stalled in the Senate; it was not brought to a vote until June 24, when it was soundly defeated by a vote of 43–32. The foreign policy concern of the State Department trumped the domestic pressure from coastal lawmakers. The most the industry could get were orders to the U.S. Tariff Commission and the Department of the Interior to conduct investigations into the impact of tuna imports on the domestic industry.[42] In the meantime, Japanese tuna continued to pour off the shelves of American grocery stores.

For the domestic tuna industry, the tariff problem was firmly lodged in foreign policy considerations associated with Iceland, Japan, Peru, and Chile. All four countries opposed increasing the tariff on tuna products. The industry wanted Congress to place all categories of tuna under the same tariff, but Congress was unable to act because an article in the Icelandic Trade Agreement required Iceland to agree to the withdrawal of an item. If the U.S. acted unilaterally, Iceland could terminate the agreement.[43] The State Department did not want to increase a tariff on a wartime ally, or create an opportunity to revisit the air base agreement. A new duty might cause unemployment and social unrest in Latin America. The U.S. was already concerned about the potential for communism to expand in Central and South America.[44]

There was no tariff on frozen tuna because it hadn't existed as a viable product in 1934 when the tariff rates were initially established. In 1946, Japan exported 4.1 million pounds of frozen tuna into the U.S. The amount doubled in 1947, and doubled again in 1948, and by 1950, frozen tuna imports were at 56.7 million pounds.[45] At least 30 percent of the American market was being supplied by Japanese tuna.[46]

The volume of imports had increased so rapidly that the Southern California fleet had little time to respond. Boats were tied up as the price of albacore dropped from $350 a ton to $300.[47] The Southern California industry began 1951 with an inventory of three million cases of tuna, depressing the market throughout the year and cutting down on the amount of fish the canners bought from fishermen.[48]

The Latin Americans were steadily increasing the fees they charged boats to pass through their waters, increasing expenses for the San Diego fleet at a time when its fish was not competitive in its own markets. In February of 1951, Ecuador increased its fees, raising the cost of fishing permits from $7.50 per ton to $12 per ton.[49] It raised fees for bait boats a month later. Permits for tuna now cost $200 and had to be used within 100 days, or a second permit was needed.[50] Mexico also increased its fees for taking bait, bringing the amount Americans paid to the government to about $330,000 annually.[51] Panama followed suit in December of 1952, pegging license costs to boat size and sharply increasing fees.[52] Each fee increase made it more difficult for American boats to compete with the Japanese tuna.

Peru had supplied the American market with canned bonito, a relative of tuna, during the war. The government was unhappy with the prospect of a tariff on the fish the Americans had welcomed so eagerly. Even three cents a pound could price Peruvian tuna out of the American market. The chief of the U.S. Fishery Mission to Peru, Robert O. Smith, wrote to Chapman in late November of 1951, pointing out that Peru could not compete with Japan; its boats were small and could not fish offshore, while its processing industry had begun in 1940 and did not achieve efficiency until 1946. The fishery was seasonal, with high production costs, and processing capacity was less than 1,000 cases daily. Most of the production was bonito, which are smaller than albacore, so handling and processing costs were higher. It cost $90 a ton to ship bonito to New York. The U.S. was the only market for Peruvian bonito. During the war, Peru had severed relations with Germany and Japan and its navy maintained a submarine patrol of its coast. "It is logical for Peru to expect somewhat more friendly treatment than would be accorded to a former enemy," Smith wrote.[53]

Now explicitly representing the tuna industry, Chapman responded that increasing the tariff was imperative if the American industry was going to survive. Peru was going to get caught in the cross fire, "but I do not see a single thing we can do about it. The industry is now quite truly fighting for its very life. It does not dare to do anything less than its utmost to get this tariff program across, and it cannot concern itself with the minor effect, in the overall picture, of our relations with Peru."[54]

The troubles multiplied in June of 1951, when the Mexican government announced it would abrogate its 1943 trade agreement with the U.S., which had established a 22 percent tariff on bonito. There would be a six-month delay for the trade agreement to be nullified, when the tariff on tuna canned in oil would go back to 45 percent. Seizing another opportunity, the Japanese escalated their imports. Imported tuna had totaled 4.5 million pounds in 1948; by 1950, it was 35.4 million pounds.[55]

By 1951, there were four different kinds of tuna products imported into the U.S., with three different tariff structures. Almost all of it was cheaper than tuna caught by American boats and canned at Southern California canneries. "These variations in duty . . . were more than merely chinks in the protective armor of the United States tariff," wrote legal scholar Harry Scheiber. "They were gaping holes through which massive imports poured into American markets."[56] In San Diego and San Pedro, the canneries were idled, the workers unemployed, the boats tied to the dock.[57]

The U.S. Tariff Commission issued its report on March 23, 1953. It declined to issue recommendations, but it drew a number of conclusions. The commission was not sympathetic. Many domestic industries had to adjust to imports from abroad. Sales of tuna were projected to continue to increase. Fishing was an entrepreneurial activity that involved greater risk than most other domestic enterprises. The economies of two allies, Japan and Peru, depended on access to the American market. The commission decided that the imports had not caused serious injury to the industry. A dissenting statement said the report should have been delayed until a more thorough investigation could be made.[58] But that was all.

A further Tariff Commission decision came in June of 1953, involving

East Coast groundfish. The New England fleet had also argued that the imports threatened the complete destruction of its industry. This time the commission recommended that tariff on imported groundfish fillets be raised from 1 7/8 cents a pound to 2 1/2 cents.[59] There was a storm of protest from the allies.

The Canadian government argued that increasing the tariff could have serious consequences, not only for trade, but for the overall relationship between the two countries.[60] The Foreign Operations Administration weighed in, warning that the operation of U.S. bases in Iceland depended on the goodwill of the Icelandic fishing communities; Iceland had recently signed a trade agreement with the Soviet Union. The State Department said there was no justification for a quota or a tariff. Either action would "conflict with our military security system to some extent in Norway, more in Canada, and worst of all in Iceland . . . As one Icelandic official has put it, in commenting on the Tariff Commission's proposal, 'Iceland cannot live on military agreements alone.' "[61]

The final decision was made by President Dwight Eisenhower; he declined to increase the tariff. While he recognized the pain, he thought expanding markets for fish products was the way to solve the industry's problems. Since the Tariff Commission had started to study the issue, the demand for fish sticks had greatly increased, bringing about an increase in American per capita consumption of fish, a figure that had held between 10 and 12 pounds per person for almost 50 years. Increasing the tariff would reduce the supply of raw material for fish sticks, hampering the development of this promising new market. Expanding the market for fish sticks "appears to hold the best prospect for a vigorous, healthy domestic industry that also best serves our international relations."[62] At the same time, Eisenhower pledged additional assistance to the industry, with research in fishing technology, conservation, and marketing.

Fish sticks were indeed popular. Demand was so great during 1953 that producers were unable to keep up with orders and a number of new companies started making the precooked, breaded food. During the first quarter of 1954, production rose to 9 million pounds and the U.S. Fish and Wildlife Service estimated that output would exceed 40 million pounds. "Consumer acceptance of fish sticks has been so widespread that some sources expect them to do for the fishing industry what fruit juice concentrates have done for the citrus fruit trade."[63]

With tariff relief opposed by the State Department on national security grounds, and President Eisenhower refusing to act, it was a difficult time for the industry. The American Tuna Association ran out of money in 1953.

Fishermen, processors, and the unions formed a new group, the Tuna Council of the Americas, to deal with the ongoing crisis in Latin America.[64]

The New England industry went back to the Tariff Commission in 1956. The commission recommended relief for the New England fleet. But once again, the State Department refused to agree. Foreign policy considerations and a desire to encourage trade relations in Iceland, in hopes of steering it away from selling fish to the Soviet Union, were more important than domestic dislocations. According to the International Cooperation Administration, a 50 percent tariff on fish would "strengthen those elements in Iceland which wish to drive out U.S. NATO troops. As fish goes, so goes Iceland."[65]

The California industry was a little luckier this time. The Tariff Commission recommended relief on tuna and President Eisenhower issued a proclamation doubling the duty on tuna canned in brine, from 12.5 percent to 25 percent ad valorem, whenever imports exceeded 20 percent of the previous year's U.S. pack of canned tuna of all varieties. The proclamation followed an agreement by Iceland to withdraw tuna canned in brine from the 1943 trade agreement, and an invocation of the rights reserved by the U.S. in the GATT agreement.[66] It was too little, too late.

By June of 1956, the White House announced a package of measures designed to help the fishing industry. There was an administrative reorganization, and the U.S. Bureau of Commercial Fisheries was created to take over the responsibilities of the U.S. Fish and Wildlife Service. The new service set up a broader research program to investigate all phases of the industry and develop new resources, improve the efficiency of fishing, and promote new fishery commodities. A $10 million loan fund was set up to make loans for maintenance, repairs, and equipment of boats, at a 3 percent interest rate.[67] A report to Eisenhower in January of 1957 said 88 applications had been made for loans, but the amounts only totaled $2.5 million.[68]

A review of tariff documents during the Eisenhower administration reveals that Eisenhower took the entire issue seriously, regularly demanding updates on the fisheries situation from aides. American policy was strongly tilted toward free trade and Eisenhower was conscious that his decisions had to be what was best for the security of the entire nation, not just fishermen. Between 1948 and 1955, there were 59 applications for relief under the escape clause and Eisenhower only granted five of them.[69]

Chapman thought the industry had fought as well as it could, but it was up against the Japanese government, not just the Japanese fishing industry. There was no choice but to continue to seek help from Congress. "Accordingly with some reluctance and trepidation, we are turning again to the Congress and the Tariff Commission. We are not sanguine. We have been along these trails before and know them pretty well. But we do not propose to give up our business without letting the Japanese know they have been in a tussle."[70]

There was a final bill, introduced by Senator Magnuson in the summer of 1957, that would provide quotas for tuna imports. The industry couldn't get the bill out of committee. "The effect of this abject defeat upon our members was profound as it was upon those of us who had engaged in the campaign, including, in particular, myself," wrote Chapman. "We not only were broke, but we were in the hole. It was quite obvious to all hands that had we not undertaken our campaign, we would not have been so broke and we would have been about where we were. About all we had appeared to do was excite a tremendous amount of antagonism against us which was in more cases than one, personal and bitter."[71]

CHAPTER SIX

Treaties

The increasing importance of redfish in northern fisheries in recent years has intensified the studies of this highly interesting fish.
—E. J. Surkova, 1959[1]

The rapid rebuilding of the Japanese fishing fleet was one of the great successes of the American occupation. Japan still had enormous shipyard capacity and once SCAP started to turn over material and supplies, recovery was rapid. The prewar Japanese fleet had been the world's largest, harvesting between 3.5 million and 4.5 million metric tons of fish a year. The U.S. catch, even pushed upward by Alaskan salmon, was less than 2.5 million metric tons.[2] All the new boats—as well as the boats of the old fleet that survived the American bombing—were confined to a tight square around the four main islands, called the MacArthur Zone, after General Douglas MacArthur. As fishing intensified, catches dwindled, and fishermen drifted out of the zone, often into the hands of Soviet naval vessels. Each arrest proved to Japan's critics that its fishermen were little better than pirates who ignored international law and could not be trusted back in their waters.

The postwar economy was severely constrained; any industry that might have a military application was banned, including the aircraft industry, steel mills, and synthetics, as well as chemicals and machine tools.[3] Once again, Japan's deep knowledge of the oceans and its fish stocks would be central to the recovery of the nation. Fuel prices were low and under international law it was possible to fish anywhere in the world, outside of a state's three-mile boundary. It was a relatively simple decision to throw the dice and rebuild the fishing fleets. Then it became very important to reclaim as many historical fishing privileges as possible and not to cede a single milliliter of water.

The Soviet Union, China, Korea, and the Philippines, as well as Australia and New Zealand, watched apprehensively as the Japanese fishing fleet was swiftly rebuilt beyond prewar capacity. The tensions of the 1930s were re-created. Neighboring states opposed any expansion of Japanese fishing territory, concerned over their security in allowing the Japanese back into their waters.[4] If the Japanese returned, it would retard the development of their domestic fisheries. The Koreans were claiming a 60-mile territorial sea and regularly seizing Japanese boats.[5] The Soviets claimed a 12-mile jurisdiction and also seized Japanese fishing vessels. The salmon industry of the Pacific Northwest also watched uneasily, bent on trying to prevent the return of Japanese boats to Bristol Bay.

There was little choice for Japan but to go north and reclaim as much as possible of the rich salmon and king crab resources that had been so profitable for the nation's motherships during the 1930s. They would build two fleets: one for salmon and king crab in the North Pacific, and a tuna fleet that would expand south throughout the Pacific and into the Atlantic and Indian Oceans. No fishermen knew more about how to catch high-seas tuna than the Japanese. They wanted the fish and they also needed jobs—with the eviction of Japanese fishermen from Sakhalin and Kamchatka, there were scores of fishermen in Northern Japan who had depended on the fisheries for at least a part of their yearly income.[6]

With the signing of the Japanese Peace Treaty in San Francisco in September of 1951, the six long years of the American occupation finally ended. The treaty was signed by 49 nations, but it was basically an American document that reflected Cold War considerations and American interests. It did not establish peace between Japan and its immediate neighbors, especially the Soviet Union and China.[7] The U.S. wanted Japan to be independent, but independence in line with American strategic interests. Scholars have suggested that the peace treaty was riddled with inequalities.[8] It certainly shaped Japan's patterns of interactions with the international community, reorienting the economy away from Asia and toward the West. The Americans pushed for control on trade with the Communist bloc and restricted Japanese imports into the American market, with the exception of tuna.[9]

In order to get a treaty, the issue of where the Japanese would be able to fish had to be set aside for later negotiation. The fishing issues were still too thorny a problem to be included in the peace treaty. The Southern California tuna fleet was in a pitched battle with Congress over a tariff on the rapidly increasing tuna imports. And the Northwest salmon industry was poised for rebellion if a Japanese boat was allowed back into Bristol Bay. The Japanese had salmon and king crab fleets ready to sail as the peace treaty was being signed. William Herrington, now the fisheries advisor in the State Department, advised patience, concerned that the appearance of Japanese boats could enrage the American fishing industry.

With a peace treaty finally signed, it was time to negotiate another treaty—this time over fish—with Japan, Canada, and the United States. The Japanese government lost no time in seeking to secure one of the rights of the Potsdam Declaration: access to raw materials that it had before the war. The Japanese had pioneered fishing in many parts of the Pacific and its industry was anxious to return and reestablish its historic right to fish beyond three miles. That certainly meant the waters of Bristol Bay. How far could the Americans keep the Japanese from their salmon stocks? How far did Bristol Bay salmon migrate into the North Pacific? It was an extremely important question.

The Americans pushed for a political line, set at 175° W longitude, and they did not want the Japanese to fish east of the line, at least for salmon and halibut. The U.S. asked Japan to "abstain" from fishing for salmon and halibut in the waters adjacent to Bristol Bay—but it could not keep them away from the king crab fishery they had developed in the 1930s. The International Convention for the High Seas Fisheries of the North Pacific (often referred to as the Tripartite Treaty) was signed May 9, 1952, at Tokyo, by representatives of Japan, Canada, and the United States. The treaty created the International Convention for the High Seas Fisheries in the North Pacific, or INPFC.

The fleet had waited a year to fish. Three days after the North Pacific treaty was signed, on May 12, 1952, the first three Japanese mothership fleets—with 50 catcher boats and 12 scouting vessels—left for the North Pacific to set miles of floating drift nets for salmon.[10] They were not allowed to fish for salmon in Bristol Bay, but there was nothing in the treaty to prevent them from catching king crab in the bay, as they had done for at least a decade.

The Japanese fleet intended to harvest as many salmon as it could on the high seas, before the fish returned to the coast to spawn. They hoped for 1.8 million salmon. The actual catch was 2.1 million fish. The 1953 fleet was even larger: six motherships, 200 catcher boats, a whaling expedition, and a king crab expedition, as well as five trawlers and a fish meal factory.[11] Despite the restrictions on where the Japanese could fish, their salmon catch reached more 163,000 metric tons in 1955.[12] The following year, there were 16 fleets fishing with 500 smaller boats.

They sent a king crab mothership in 1953, with six trawlers and six deck-loaded fishing skiffs. They hoped to pack 50,000 cases of crab but managed

58,240. Trawlers located the crab grounds while the smaller fishing boats set the tangle nets, gear that would later be banned for its bycatch of other species. The pack came from the catch of an estimated 948,842 king crab. Within a decade, there were reports that the fishing was not as productive and the crabs were smaller.[13]

The Tsushima Plan, started in 1954, was aimed at moving boats from the crowded local grounds into the new salmon mothership fishery in the North Pacific, or into the tuna fleets that had begun to expand into the Eastern Pacific and the Indian Ocean.[14] It called for rapid development of Japanese enterprises and worldwide exploitation of the oceans. The government appropriated 550 million yen for overseas development for 1956, 1957, and 1958. It was increased to 980 million yen in 1959. There was additional money to fund overseas bases and to build and maintain a fleet of research vessels, so fishing technologies could be adapted to different conditions in the Pacific, Atlantic, and Indian Oceans.[15]

The Japanese Fisheries Agency announced a five-year plan to build tuna vessels, 230 steel-hulled and 1,280 wooden-hulled vessels, between 1954 and 1958. The expansion was expected to double productive capacity and would accelerate the conversion of vessels of various classes to expand the Japanese skipjack fleet.[16] The whaling fleet had been rebuilt, starting in 1949 with two new floating factories. By 1960, they had seven fleets working in the Antarctic.[17]

The American Embassy in Tokyo reported in early 1953 that the government had given "considerable" aid to the fishing industry during the previous year. There were multiple projects to build boats.[18] The subsidies included 262 million yen, or $730,000, for biological and oceanographic studies of tuna and tuna-like fisheries. There were funds to operate 14 research vessels, four of which were primarily dedicated to researching tuna.[19]

For the 1955 season, the government announced it would increase its fishing in the Aleutian area of the North Pacific by more than 63 percent, to 334 fishing trawlers. There would be 11 mothership operations and expanded canning facilities.[20] Japan began negotiations with the Soviet Union to restore prewar access to the waters off Kamchatka and for resuming fishing for salmon, trout, and crab in the Sea of Okhotsk.[21] Japan was once again substantially enlarging its high-seas fisheries.

The key to Japanese dominance of the markets was the low wages it paid its fishermen. Large Japanese fishing companies would only guarantee a minimum monthly wage of 5,000 to 6,000 yen, or about $13.90 to $16.50 in American dollars.[22] The work was arduous and conditions were harsh. The working day would begin early in the morning and last until the catch was processed. Vessels at sea had quotas they needed to meet. Conditions were just as bad at the onshore canneries in the Northern Kuril Islands, with 21-hour workdays and poor food and sanitation conditions. The wage structure emphasized bonuses for increased production, encouraging workers to remain on the job even when they were clearly too tired to continue.[23] There was little change from the conditions Takiji Kasahara had written about in the late 1920s.

The rapid escalation of the Japanese high-seas drift net fishery for salmon prompted the Soviets to establish an arbitrary border, the Bulganin Line, enclosing all of the Sea of Okhotsk and much of the area east of Kamchatka and the Kuril Islands. They demanded that the Japanese apply for permission to fish in their waters. The two sides quickly negotiated the Japanese Soviet fisheries agreement, which set annual quotas for salmon, herring, and crab.[24] It set vessel limits, regulation of placement of nets, minimum net gauge, and a minimum size for catching herring and crab.[25] The agreement also contained what would become a very providential clause, allowing the Japanese future access to a fish called walleye pollock (*Theragra chalcogramma*).[26] The quick negotiations allowed both countries to get back to what was important: catching fish.

The Japanese were also anxious to rebuild their whaling fleets. Limited whaling had been allowed under the occupation, but in 1951, the Japanese replaced two smaller floating factory whale ships with the *Tonan Maru* and the *Nisshin Maru*.[27] They bought two more whaling ships, the *Thorshovdi* from Norway and the *Southern Venturer* from the United Kingdom, allowing the Japanese to increase their share of the annual Antarctic whale catch quota from 41 percent to 50 percent. For the 1962–63 seasons, they operated seven whale factory ships in the Antarctic grounds.[28]

From 1956 until 1973, the fastest-growing share of the total catch came from the distant-water fisheries, especially in the North Pacific Ocean. There were no restrictions on access, fuel was still inexpensive, and the continued increase in the catch all favored further investment into the distant-water fleets.[29] There was also the knowledge that two decades of research by Japanese fishing boats and educational training vessels had accumulated as they fished the eastern Bering Sea shelf. The shelf is about 1,200 km long and 500 km wide at its narrowest point, and it is the widest continental shelf outside the Arctic Ocean. It is essentially a large, featureless plain that deepens gradually from the shore to about 170 m at the shelf break, and it is a superb place to grow fish. The Japanese resumed fishing there in 1954, with two mothership fleets. By 1961, they had 33 fleets, with a total of 377 affiliated trawlers. These fleets consisted of frozen fillet processors and fishmeal processors as their motherships, with flounder (*Paralichthys olivaceus*) as the primary target; in 1961, flounder accounted for three-quarters of the catch.[30] The stock was quickly depleted.

Another of the fish that grew well on the shelf was yellowfin sole, *Limanda aspera*, a flatfish that lives on the soft, sandy bottom. Its native habitat is the temperate waters of the Pacific, from Korea and the Sea of Japan and Sea of Okhotsk, to the Bering Sea and south along the West Coast of Canada. The mothership fleets produced frozen fillets and the waste was turned into fish meal. By 1961, there were five meal ships, 18 freezer motherships, and 300 catcher vessels. That was the year the yellowfin sole fishery peaked, two years after it started. Catches declined rapidly after that.[31]

There were other fish on the shelf, including walleye pollock. Pollock are found around the entire North Pacific rim, from the waters of the Sea of Japan, to the Sea of Okhotsk, and into the Bering Sea. They are a relatively new species, only around for 3 million years, and they are closely related to Atlantic cod. They occur in large dense schools off the sea floor, which means they are easily caught. They also have a low bycatch of other species, making them easy to sort and process.[32] The flesh is white, with a low fat content, so it is good for both fillets and minced for fish sticks and fish burgers. There was a long history of harvesting pollock in both Japan and Korea. The initial catch in 1960 was just 25,000 metric tons. The Soviet boats joined the fishery in 1959 and catches exploded to 655,800 metric tons by 1969.[33]

Fueling the growth, especially on the Japanese side, was the discovery that pollock was the perfect fish to be turned into a fish paste called surimi, and surimi could be frozen without damage. Surimi products could now be made available all year long in Japan and demand soared. Within 13 years after the introduction of frozen surimi in 1960, the Japanese surimi-based product industry doubled in size. The frozen surimi industry by the mid-1970s was producing 355,000 metric tons annually. As of 1984, the surimi industry was a $500 million business in Japan.[34] Though almost any fish can be used to make surimi or fish paste, pollock offered an unmatched combination: they were abundant and easy to catch, making them economical for an industrial fishing operation.

This huge expansion in the number of massive motherships required a large number of trawlers to supply them with fish. More big, deep-sea trawlers were built in 1959, to fish in the North Pacific.[35] The boats were also sent offshore, in the South Pacific off Australia and New Zealand, and off the West Coast of Africa. Part of the reason for the expansion was the increasing restrictions on the North Pacific salmon fishery, with the U.S.,

Canada, and the Soviets all concerned with limiting the Japanese high-seas fishery so more salmon would return to spawn.[36]

The Japanese also went south in pursuit of tuna. During the 1920s, they developed a lucrative fishery in the waters of their former island possessions, the Marshall, Mariana, and Caroline Islands. The islands were now controlled by the Americans, but thanks to the new tariff regulations, Japanese boats could deliver to the new Van Camp processing plant in American Samoa. Fish processed in the islands could enter the American market at no tariff.[37] The duty on tuna canned in oil was 35 percent; for fish canned in brine or water, it was from 6 to 12.5 percent. Starting in January of 1954, StarKist bought tuna from seven Japanese tuna boats. The landings grew quickly, especially when boats from the Republic of Korea and Communist China began to land tuna at the American processing plants.[38]

Japan launched the 600-ton tuna "guidance ship," the *Sagami Maru*, in late 1956. It was owned by the Kanagawa prefectural government and would conduct exploratory fisheries for tuna, first in the Indian Ocean off the northern tip of Madagascar and Seychelles. It would be the first Japanese ship to pass through the Suez Canal, delivering tuna to Genoa for packing.[39] A fleet of long-range tuna vessels were built for the Indian Ocean fishery, using vessel designs from the Tokaiku Marine Research Station at Tokyo's modern vessel testing facilities. "The great long-liners are being built primarily for use in the rapidly developing Indian Ocean fishery, which has proved astonishingly productive," reported *Pacific Fisherman*.

A radio station kept in contact with all the longline vessels across the world, collecting and distributing charters on catches and water temperature. Monthly charts showed the average catches and number of vessels in each sector. Boats fished at night, as fishermen searched for plankton that fluoresced, attracting the fish. They soon learned that bluefin tuna passed the nights at the surface of the ocean.[40]

There were Japanese boats off Argentina and Brazil in 1955. They also had big factory processing boats operating around Australia and New Zealand, in the Indian Ocean, off northwest Africa, in Newfoundland, and in Antarctica. Ships were often at sea for a year and cargo was shipped back to Japan. The boats targeted highly prized fish such as sea bream, croakers, shrimp, prawns, and squid.[41] In particular, they were after tuna, especially bluefin.

The Japanese fishery began in the middle of 1956 in the waters around the boundary between Venezuela and Brazil. They were targeting yellowfin and albacore to can when they encountered bluefin tuna. The longline fishery would set nets more than 150 km long, fitted with 3,000 hooks. The hooks had to be baited by hand, making the fishery very labor-intensive. Within a very few years, the catch of bluefin peaked at about 13,000 metric tons in 1964. A few years later, almost no bluefin tuna were caught in the Southern Atlantic.[42]

At home in Japan, as the Korean War escalated, the domestic economy was improving. With more fish flowing into Japan, seafood consumption increased, creating a demand for sushi and sashimi. The industry heavily promoted sashimi made from frozen tuna at a time of increasing incomes in Japan.[43]

An American food scientist at Michigan State University, Georg Borgstrom, was one of the first to realize just how rapidly and extensively the Japanese had rebuilt its fishing industry. He published a book in 1964 alleging that Japan's codevelopment work and joint ventures with other countries dwarfed the efforts being made by the U.S. Food and Agricultural Organization. He listed six different kinds of agreements in use: joint fishing and processing companies, contracts to supply local markets, technical assistance, exploratory fishing, transit bases, and selling directly in foreign ports. The Japanese had cooperative operations in 50 countries. "God said, according to Genesis, 'Let them have dominion over the fish of the sea,'" Borgstrom wrote. "Why have we allowed this commandment of the Lord to be fulfilled by the Soviets and the Japanese?"[44]

By the 1960s, Japan had engaged in 13 contracts with countries in Europe, Asia, and Oceania. A decade later, there were 77 agreements, including those in Africa and Latin America. The estimate for 2011 was 115.[45] According to Borgstrom, the Japanese arrived at an agreement with Brazil in 1952, and began to map the waters of both Brazil and Argentina. Between 1955 and 1959, nine agreements were signed between Japanese fishing companies and the Brazilian government. Two joint venture companies were created and a third was planned. Fishing rights were granted to 20 Japanese fishing boats, while a Japanese-Brazilian company engaged in coastal whaling. "The Japanese are largely responsible, though not always directly, for the rapid growth of the fish industry in southern Brazil."[46]

The tuna longline fleet saw dramatic expansion. Landings increased by a 3.9-fold increase in just a decade, as the tuna fishery expanded from the Pacific into the Indian, and then into the Atlantic Ocean.[47] Within a few years, the European market for tuna was glutted and the Japanese began to dump tuna into the American market.

"This caused the American market to glut and the price to break," Wilbert Chapman wrote. "This was the final straw that broke the San Diego bait-boat fleet, the American Tuna Association, a number of fishermen and boat owners, and caused the whole southern California fishing industry to get sufficiently excited so that for a period of about five months the canners, unions, and boat owners practically worked together (the longest such united frenzy so far in the history of the industry)."[48]

THE UNITED STATES AND THE SOVIET UNION

A massive flotilla of Soviet fishing boats appeared in the water off Western Alaska in March of 1959. The *Seattle Times* flew over the boats, took pictures, and went looking for somebody who could explain what the boats were doing. Most of the boats were about 125 feet long; some were more than 300 feet long. They found a Seattle biologist, Dayton Lee Alverson, who knew exactly what the boats were and what they meant. And by this time, he also had a pretty good idea of how many fish the Soviets were going to catch, because for the last decade—since his days as a fishery scientist at Astoria in 1947—Alverson had been studying the astonishingly diverse world of rosefish and its cousins.

The U.S. Fish and Wildlife Service had created its exploratory gear base and found its Pacific research vessel, the *John N. Cobb*. Its mission was to help fishermen find new Pacific resources. It was exciting work. On one of its early voyages, the *Cobb* found a 27-million-year-old submerged basalt volcano about 280 miles west of Gray's Harbor on the southern Washington coast. The base of the now-named Cobb Seamount is 9,000 feet deep and it rises to within 108 feet of the surface, providing a large area penetrated by sunlight and supporting a complex ecosystem of marine species, dominated by rockfish. Seamounts can trap strong eddies of water, sheltering fish larvae that otherwise would be swept into the open ocean, promoting better survival rates.[49]

There were other productive areas as well, in deeper water and on the slope of the continental shelf, where fishermen were finding large numbers of the bright red fish they called rosefish, or Rosies. And what was even more astonishing was the large number of small fish, what scientists call a cohort or a year class. "The enormous schools of adult Pacific ocean perch that fishermen were seeing on their echo sounders represented the accumulated net production of over eight decades," wrote Donald Gunderson, one of the first American scientists to study rockfish.[50]

There are some 70 species of *Sebastes* rockfish, with distinct populations found in each tropic zone. There are blue and black rockfish, as well as canary rock, in the waters of the kelp zone. The midcontinental shelf is home to immense schools of canary, widows, and yellowtail rockfish.

At 600 feet there are Pacific ocean perch. Deeper still, there are shortraker and rougheye. In waters of 3,000 feet there are thornyheads. Some species can live up to 200 years. Females bear live young, which might be an evolutionary adaptation to the brief and unpredictable spring plankton bloom in northern waters.[51] When spawning events are uncertain, multiple year classes help buffer the disturbances on the overall population.

The *Cobb*'s mission changed abruptly in 1954, after the signing of the peace treaty with Japan and the Tripartite Treaty. When the Americans negotiated the fisheries treaty, they had very little information on how far west salmon from Bristol Bay migrated before they returned to their home streams to spawn. There was an urgent need to find out where the fish migrated—did the new political line at 175° W protect American salmon from the ravages of the Japanese fleets? The *Cobb* was sent to the North Pacific to help determine where salmon migrated. It was not what Alverson wanted to do, and he went to work for the Washington State Department of Fish and Game, which allowed him to continue his trawl development work, including monthly columns in the *Fisherman's News*. When the *Cobb*'s mission changed back to more experimental fishing in 1959, Alverson returned to the U.S. Fish and Wildlife Service.[52]

Two years earlier, Alverson had been in Hamburg, Germany, for a Food and Agricultural Organization of the United Nations Conference. He saw fishing boats as big as cruise ships, with big diesel engines, powered deck machinery, and sophisticated new navigation technology. They dwarfed the small trawl boats Alverson worked with in Washington and Oregon. "Within a decade the new technology and the global demand for animal protein would foster a modern high-seas fishing fleet what would invade every major ocean of the world, from pole to pole . . . I was almost embarrassed to give my talk on small West Coast stern trawlers," Alverson later wrote in his memoir.[53]

Fishing was being revolutionized and food from the sea was playing an increasing role in the economies of Eastern Europe, Japan, and Korea, and especially China.

> The conference participants were so enthusiastic about the growth of the ocean fisheries that they gave the impression of being caught up in a fever . . . What was a surprise, however, was the sudden emergence of the eastern bloc nations, which had no substantial history in world fisheries prior to World War II. The Soviets talked about their ten-year plan and their intent to greatly expand high-seas fishing using mother-ship operations and a fleet of modern, self-contained catcher/processor ships. They

also talked about their efforts to learn more about the oceans' great capacity to produce food for mankind. Most Americans were unaware of, and/or, took little interest in, the race for the sea.[54]

Bob Hitz, who began his graduate work at the College of Fisheries in 1958, remembers Alverson coming to speak to his class after he returned from Hamburg. Alverson described the distant-water trawling fleets he had seen. Hitz interviewed for a job on the *Cobb*, and Alverson "told [him] that the rockfish were a large group in the Pacific and he believed that one species, the Pacific ocean perch, or Rosefish, would dominate the commercial catch in the future."[55]

It must have been with deep misgivings that Alverson identified the Soviet fishing boats. The Japanese had been fishing in the Bering Sea since the early 1950s, but these new Soviet vessels "raised the hair on the necks of fishermen from California to Alaska," Alverson wrote, adding that several scientists didn't believe the Bering Sea could sustain such a large fishing fleet. And indeed, some species could not.

The Soviets had done exploratory work near the Pribilof Islands during the 1930s and found large catches of rockfish. When fishing resumed in 1958, about 90 percent of the catch was *S. alutus*, or Pacific ocean perch. The Soviets carried out extensive investigations in the eastern Bering Sea before they began fishing, looking for information about the distribution and abundance of the fish, as well as currents, and the topography and geomorphology of the sea floor, typical of the investigation the Soviets did for all their global fishing operations. The Far East Fishing Company began catching perch in May of 1960, with 25 vessels using a variety of nets. The following year, the floating cannery, the *Andrei Sakharov* built in Leningrad, was sent to the Bering Sea. More than 15,000 GRT, it was sheathed for ice and equipped with a machine shop to help repair other vessels.

The Soviets moved from the Bering Sea in 1959 to the Gulf of Alaska in 1962, to the Aleutian Islands in 1963, and finally to waters off Washington, Oregon, and California in 1966. The fishery "continued east as one stock after another was depleted," Donald Gunderson, a retired University of Washington fisheries scientist, wrote in his 2011 memoir. "Washington and Oregon were their last stop."[56]

The Soviets found that all along the North Pacific shelf, for thousands of kilometers, at depths between 200 and 450 meters, and stretching from the Aleutian Islands all the way south to California, there were schools of brightly colored rockfish. They were plentiful on the rocky reefs near the deepest margins of the continental shelf, waters that most coastal fishermen could not access.[57] The ships caught an average of seven to nine tons of fish per hour, with some hauls as large as 16 to 18 tons. The Pacific catches made up more than a quarter of the total Soviet catch between 1970 and 1974.[58]

By 1963, the Soviets were fishing year-round along much of Alaska's coastline for rockfish and herring. They followed the rockfish south to Washington and Oregon in 1966 but catches dropped quickly. They shifted to a large-scale fishery for Pacific hake, *Merluccius productus*. There was no American market for the soft-bodied hake. The fish would spawn off California then migrate north, growing and putting on weight. The Soviets moved south in the spring, then followed the hake to British Columbian waters.[59]

The rosefish catches peaked off Alaska in 1965, when 15,600 metric tons were caught. Catches dropped rapidly to 6,000 metric tons by 1968.[60] The boom was over, and American biologists such as Gunderson were just starting to study the bright red fish. Pacific ocean perch were considered "depleted" off Oregon and Washington by 1969.

While the Soviets fished, American fisheries officials began to systematically study the new boats in their waters. Hitz recalled flying in a twin-engine Piper Aztec over the Soviet fleet in 1967, photographing each vessel and recording its name and position at sighting: "When we spotted a vessel we would drop rapidly, losing altitude until we were just above the water's surface. The plane cruised at about 180 knots and as we neared the vessel our pilot lowered the flaps, slowing down to about 110 knots, and as we passed Brad read the numbers and name and recorded them in his notes. It was exciting, a lifetime event and one I will never forget. We did it time after time and at least 30 vessels were counted during the trip that day."[61]

Between 1960 and 1975, the Soviets greatly increased their fleet and the total tonnage they caught. The annual production of each worker tripled.[62] Fish was making up more than a third of the total animal protein produced in the Soviet Union during 1961. Pacific fisheries set new records and exceeded the initial target catches, making it easy for A. A. Ishkov—the minster who would head the Soviet fisheries ministry for 25 years—to get the backing of the Central Committee of the Communist Party to continue to pour resources into increasing the fishery. In a speech to fishery workers on February 14, 1961, Ishkov said that with Soviet fishing well established in the Atlantic, it was time to turn the country's attention to the Pacific, to broaden the fishery in the Bering Sea "in every possible way," not only for flatfish and rockfish, but for salmon, halibut, and black cod. This expansion would take two directions: factory ships for catching and processing, and mother ships for taking catches back to port and resupplying the fleet.

The Soviet boats were catching fish but they were inefficient and not meeting demand. "Thc industry [did] not study and take advantage of the experiences of the leading vessels," so productivity was not increasing as quickly as it could. Too much time was lost in port doing repair work. By January of 1963, Ishkov said there would be a single system for planning fishing operations of all types of large fishing vessels. "At present time, investigation of the fish supply in the open seas and oceans falls behind the needs of the fishing fleet. It should surpass it."[63]

Ishkov also announced excellent results from the fourth year of the scvcn-ycar plan. The target for 1962 was met on December 15, 1962, and an additional 200,000 metric tons was added to the quota. The modern trawler should catch from 5,500 to 6,000 metric tons of fish each year. If the transfer of the catch at sea could be streamlined, the average catch of each trawler could be increased to 7,500 or 8,000 metric tons each year. More fish were available and fishing could expand.[64]

The Soviets expanded into whaling in 1959, building two enormous whaling ships of 45,000 GRT each, and several small modern whaling factories. The fleets operated in the Antarctic, replacing Norwegians, British, and Dutch companies.[65] There was also an armada of 67 whale-catcher boats, said to be capable of running at 19 knots.[66] They also took over the Japanese processing equipment that had been left on Sakhalin Island and Kamchatka. They were anxious to get back to producing canned king crab, as they had done before the war. They also deployed the American Lend-Lease vessel, the *Alma-Ata*, according to several items in *Pacific Fisherman*, cited by biologist Robert S. Otto in his history of the development of the king crab fishery in the Bering Sea.[67]

The Soviets set up an extensive network of agreements that authorized operational bases and joint ventures for fishing and processing. The agreements secured access to fishing grounds and more than 2 million tons of fish, most of it transferred back to the Soviet Union for processing. The Soviets concluded trade agreements with Ghana, Sudan, and Somalia, as well as formal technical assistance programs for Cuba, Egypt, Indonesia, Yemen, and Guinea. "Like many other nations, her motives are humanitarian, economic, and political in these aid programs."[68]

As the Soviets expanded fishing, they also joined management organizations, such as the International Whaling Commission (IWC) and the International Council for the Exploration of the Seas (ICES). When they began to fish off Newfoundland, they joined the International Commission for

North Atlantic Fisheries (ICNAF). When they began to harvest fur seals in the North Pacific, they renegotiated a new Pacific fur seal treaty in 1957. The Soviets never "sought structural changes in treaties and abided by mesh size and size limits to protect younger immature fish," according to a 1968 analysis by two American scientists.[69] Joining associations, abiding by the rules (at least publically), as well as the research the fleets engaged in, helped to promote the modernity of Soviet fishing and how it was grounded in fisheries science. There indeed was good fisheries and oceanography science coming from Soviet scientists, but it had little impact on a political structure that was wedded to production at all costs, randomly creating quotas and catch goals.

Soviet scientists attended a 1956 conference in Copenhagen, joining scientists from more than a dozen nations to talk about the new fisheries developing in their home countries. All across the North Atlantic, from Maine to Greenland to the Barents Sea, fishermen were fishing deeper water and finding large catches of a bright red fish that was broadly called either *Sebastes marinus* or *S. viviparous*. Trawlers from California north to British Columbia were also finding large quantities of another fish, *Sebastes alutus*, more commonly called rosefish or Rosies. Were the *Sebastes* fish all from the same taxonomy? Were some of them subspecies? Were there differences between fish caught inshore and the ones fishermen were finding at 200 fathoms (1,200 feet) of water? How fast did the fish grow? And, most important of all, how long did this medium-sized fish live? It would turn out there are dozens of different kinds of *Sebastes* rockfish, some found in shallow waters, some in deep water, almost to 10,000 feet (1,667 fathoms). The dominant species was called *Sebastes alutus*, but it had a variety of informal names. Fishermen in both oceans called them rosefish, redfish, Rosies, Atlantic perch, or red snapper.

Scientists were just starting to share their research on the fish. ICES held a conference in Biarritz in 1956 on the methodology that they should use to interpret otoliths, calcareous "ear stones" that lie on a bed of nerves deep inside the fish's head. The ear stones enable fish to orient and maintain equilibrium.[70] The otoliths had rings, but how should they be counted, and what did the rings mean? Rockfish presented complex taxonomy problems, since there were so many species with subtle variations, depending on the depth where they had been caught. But the chief difficulty with a species that had such wide distribution was to figure out how they could be conserved. And that meant scientists had to figure out how old they were, and how to read the record on the otoliths. It was difficult, because counting the rings destroyed the otoliths, making it impossible to recount and verify.

The scientists called for another meeting, focused on redfish, held in Copenhagen in 1959. Alverson attended, delivering a paper he had written with his Astoria colleague, Jergen Westrheim, "A Review of the Taxonomy and Biology of the Pacific Ocean Perch and Its Fishery." Their paper concluded that many more fish could be caught if there was sufficient market demand.

The scientists who gathered in Copenhagen in 1959 were divided about the age of rockfish. Some thought redfish grew fast, matured by the age of three or four, and died in no more than a decade. Others thought the fish matured at 10 years and could live to 30 or more years. Russian scientist E. J. Surkova, from the Polar Research Institute of Marine Fisheries and Oceanography (PINRO) in Murmansk, laid out the problem: if the fish grow quickly, the population could be hurt by overfishing—but it could be restored from other, stronger, year classes. If this was true, that the fish grew quickly, it was not as important to regulate fishing. If the fish were older, then the populations were more stable, able to withstand cycles in the ocean. But if these populations were overfished, recovery would not be quick. The stocks could "reach a state of depression, which can be restored only with difficulty."[71]

THE AMERICANS, 1956–66

The appearance of the Soviet fleet, so big, so many ships, and so close to shore, made Americans deeply uneasy. The foreign boats were accused of spying and ruining the gear set by American fishermen. Most of all, they were accused of taking all the fish. The Soviets engaged in pulse fishing: taking everything they could catch, fishing until the nets came up empty, then moving to new water to do it again. The fleets were vast, dozens of motherships and supply boats, hundreds of boats, miles of net in the water. As the technology to find fish improved, it was more like strip-mining than fishing.

The Departments of State and Defense still stoutly supported a three-mile limit, concerned that if they entirely banned fishing boats from their waters, other countries might ban American boats—or American naval ships, or submarines. This was a crisis that created multiple pressure points and brought a powerful foreign policy dictate into conflict with the small and disorganized American fishing industry, at least at first. Discontent over the Soviets boats steadily increased. "Fisheries are one of the major battlefields of the cold war," warned Washington Senator Warren Magnuson, who had been sounding the alarm over Soviet fisheries since 1951, when he complained about the U.S. government building fishing boats for the Soviets.[72] But despite the mounting concerns about Soviet fishing fleets spying on the Americans, the response to the escalation of foreign fishing was extremely fragmented and halting.

What should the Americans do? There was no overall easy objective. There were 11 agencies involved in fishing in some way, with little incentive for them to cooperate—there were not even mechanisms if they had wanted to cooperate, according to Edward Wenk Jr., a policy staffer during the Kennedy and Johnson administrations, who wrote about the messy policy debate in *The Politics of the Ocean*.[73]

The history of American maritime policy is found within a complex stew of competing agencies. The coast guard was created in 1790, the navy in 1798. Federal intervention into the fisheries was piecemeal and spasmodic, responding to crises and with no overcall attempt at an integrated policy. In 1870, the British sought damages of $14.8 million for the amount of mackerel the Americans had harvested in Canadian waters.[74] The Americans were horrified and refused to pay. But the British, being more experienced at running an empire, had the numbers, a treaty, and a tax bill. This early territorial dispute over fishing was an impetus for Congress to establish the U.S. Fish Commission in 1871. Much of the focus for the early commission was on the hopeful belief that hatcheries could replace the wild fish runs being decimated by overfishing and industrialization.[75] If the commission could just make more fish, it wouldn't have to bother with restricting fishing.

Federal fishery scientists saw their role as supporting the economic role of fishing, finding markets so that American fishermen could sell more fish and make more money. But the problem was not finding new markets; the problem was the American industry could not complete with cheap tuna from Japan and cheap cod from Canada and Iceland.

The postwar American fishing industry spent most of the 1950s fighting with the federal government for a tariff to protect their markets. Despite many hearings, they failed dismally. The next line of attack was to argue for government assistance to make them more competitive. The industry that had prided itself on free enterprise and no subsidies was now seeking a share of the support the government paid to farmers.[76] Americans were buying a lot more fish, but unfortunately it wasn't fish caught by American fishermen. In 1955, U.S. fishermen produced about 68 percent of the seafood eaten by Americans. In the space of a decade the ratio flipped, with foreign boats supplying 70 percent of the seafood eaten by Americans. Domestic production was flat, about 2.2 million tons a year.

There were more serious problems. Some 20 American tuna clippers had been seized by Peru, Ecuador, El Salvador, and Panama by 1954.[77] The industry persuaded Congress to pass the Fisherman's Protective Act, which required the State Department to ransom back fishing boats seized in Latin American waters for illegal fishing. The department did not want fishermen to buy licenses, since that implied recognition of Latin American territorial claims, something neither State nor Defense would allow. Paying the fines and securing the release of the fishermen and the boats was a small price for the government to pay for the strategic value of staking a territorial claim, even if the claim wasn't recognized and the boats kept getting arrested.[78]

The Saltonstall-Kennedy Act in 1954 collected 30 percent of gross receipts of custom duties on imported fish products and directed them to doing marketing and research for the industry. After the passage of the Fish and Wildlife Act in 1965, the bulk of the money was soon devoted to general operations of the Bureau of Commercial Fisheries (BCF).[79] The new bureau "was not much more important than the Fish and Wildlife Service had been," said Wenk.[80] Recreational fishermen and the processing industry sought to blunt the power of the commercial industry.

The new bureau focused on finding overseas markets for American fish, even though most of the traditional American fisheries were static, or, as the bureaucrats put it, "mature," including New England cod, haddock, and ocean perch, Northwest salmon, and halibut.[81] The Act also created a Fisheries Loan Fund that doled out $31.3 million between 1957 and 1973.[82] "The United States should be the leading fishing nation of the world because it has all the qualifications for this position," the new bureau stoutly argued in 1968.[83] Despite extensive coastlines, stocked with many different kinds of fish, Americans were importing more and more of the fish they ate, and the American fishing fleet, which should be the best in the world, clearly wasn't.

While most other countries created direct loans to fishermen to build boats during the 1950s and 1960s, the Americans would not do so until 1970. Fishermen had been eligible for low-interest loans since 1935 but the Eisenhower administration found that no boat loans had been made.[84] The battle over how to help the fishing industry centered on tariffs, not on assistance to the fleet to build boats. American trawl boats tended to be small, self-financed, and meant to fish in local waters; traveling to catch fish increased costs. The exception to local fishing was the California-based tuna clippers that could travel 2,000 miles to catch fish that would be frozen then transferred back to shore for processing. The Japanese also had to travel to catch fish, but they canned the fish onboard; they could deliver canned tuna to American supermarket shelves more cheaply than the Americans could. The Soviet ships fishing off the U.S. were catching and processing at sea, filleting and freezing fish, and turning the waste into meal and oil. There were no American boats that could operate at that scale.

Something needed to be done for the American fishing industry. When President John F. Kennedy took office in 1960, he promised a "new frontier" in domestic and foreign policy. The space race was fully underway, but what about the oceans? Kennedy was from Boston and well aware of the problems of the New England fleet. Fishery officials, in their new agency, were looking for a way to get noticed. They came up with an idea that the United Nations had endorsed a few years earlier, creating fish protein concentrate (FPC). It caught the attention of Kennedy's Interior Secretary, Stewart Udall. It would catch the attention of many other politicians, including Vice President Hubert Humphrey and President Lyndon B. Johnson. Proponents predicted it could help balance the diet for one billion people at a cost of a half-cent a day.[85]

If ever a situation called for a magic bullet designed to cure multiple ills, it was the ailing American fishing industry. And here was an odorless, tasteless, protein-rich powder, a fish flour that would increase the use of ocean resources, improve the economic status of American fishermen, and, perhaps most important, enhance the world image of the U.S. in helping to alleviate world hunger. The problem was that the technology had not yet been developed. And despite the money thrown at it, FPC could not be produced economically. Many other countries also tried to produce FPC during this period, but none were able to do so, at least not economically. But no country spent as much money and political capital as the U.S.

The BCF, anxious to showcase its expertise, went to work experimenting with the fish powder at its own laboratory in College Park, Maryland. There were two potentially viable extraction processes; the lab chose one developed in Canada. The inventor of the second process, Ezra Levin, president and founder of VinBio Corporation, was furious. His complaints and criticism would dog the project for the next decade.

The American dairy industry, concerned about a new product that might erode the sales of nonfat milk, prodded the Food and Drug Administration (FDA) not to approve the powder for sale, because it contained fish viscera, bones, and scales; all considered "filth." The FDA prohibited it for sale within the U.S.; the powder could, however, be sold abroad. Further, the FDA then decided it could only be sold in one-pound bags, making it unusable on an industrial scale. Then the FDA announced a restriction on how much fluoride could remain in FPC after it was processed. The only way to reduce the fluoride level was to debone the fish, greatly increasing costs.

Fish protein concentrate "illustrates the danger of exposing a new, untested technology to the political arena before its problems and potential have been adequately assessed," according to Ernst R. Pariser and Christopher J. Corkery, who detailed the entire mess in a 1978 book called *Fish Protein Concentrate: Panacea for Protein Malnutrition?* The short answer to their question was no. What they found fascinating about FPC from a policy perspective was the large number of "individuals and institutions—each with its own set of priorities and motivations—that had an impact on the character and destiny of the FPC program."[86]

In some ways, the story of FPC is the same as the story of the *Pacific Explorer* project. Both were developed under the same rationalization. The government would give some start-up money, but the projects would not be in competition with private industry. Neither project engaged in competitive bidding. The government would only assist with production development and marketing until a private company could take over.[87] It didn't work with the *Explorer* and it didn't work here. The scale of the problems—creating a tuna industry in the Western Pacific using a handful of boats, and solving world hunger—reveal how little federal officials knew about the fishing industry they were supposed to be representing, or about the fish they were supposed to be conserving.

FPC was embraced by the academic community, charged with creating and perfecting the final product. Pariser and Corkery contend the push for FPC created a technological imperative, ignoring questions of how practical the product would be. The idea was to take low-quality, cheap fish and turn them into a dried product that could be stored without refrigeration and used to alleviate malnutrition. "Nutrition activists, politicians, regulatory agencies, and private industry came into conflict in a manner most often detrimental to the implementation of FPC as a nutrition supplement," they wrote, with great restraint considering it was their displeasure with the process that prompted them to write their book.[88]

The project tried to create a universally acceptable, grayish-white, odorless, tasteless product. A spoonful in a glass of water would supplement nutritional levels in third-world countries. But the criteria created both technical and economic problems. It was a high-technology enterprise, reflecting massive amounts of capital, energy, and expertise, all of which were lacking in the developing world where it was going to be deployed. It was costly to produce, only provided a supplement to nutrition, and ignored that some developing nations had no history or culture around eating fish, and had limited access to clean water.

The Kennedy administration budgeted $50,000 to start an investigation, opening a tap that would turn into a small but focused torrent, soon consuming a big share of the money spent on fishery development. President Johnson decided FPC was part of his Great Society and the war on hunger. He directed his science council to "work with the very best talent in this nation to search out new ways to develop inexpensive, high-quality synthetic foods as dietary supplements," he announced on February 10, 1966. "A promising start has already been made in isolating protein sources from fish, which are in plentiful supply throughout the world."[89] The full-blown science-fiction version of FPC called for the technology to be placed on a ship that would sail to a country with a hunger problem, catch fish, make the powder, distribute it, then move on to the next humanitarian crisis.

Congress passed legislation to build a demonstration plant in Aberdeen, WA. With the start of construction came another round of technical delays and difficulties. By the time the plant made some powder, it cost $5.80 a pound, not the few cents its early proponents had predicted. This was a bitter failure on many fronts, not the least of which was concern over the global food supply. The President's Science Advisory Committee in June of 1966 released a report projecting a major breakdown in the world food economy within two decades, "a crisis that could be avoided by increasing the harvest from the sea."[90]

CHAPTER SEVEN

Imperialism

> The United States, however, has not been aboard this world ship of fishing progress.
>
> —Senator Warren Magnuson, 1968[1]

Iceland's unexpected ally to protect its fish stocks turned out to be the United Nations. Iceland's UN representatives in 1949 raised the issue of expanding the territorial sea and asking the International Law Commission to once again study the issue. The European countries fishing in Iceland's waters lobbied fiercely against the motion, but it passed. "By adopting this resolution, the General Assembly implicitly agreed with the Icelandic delegation that the opinion of the leading West European powers—that the 3-mile limit was firmly established rule of international law of the sea—was wrong, and accepted the necessity of studying the regime of territorial waters," wrote Hannes Jónsson.[2] It was an important milestone in the evolution of what would come to be called the Law of the Sea.

Iceland moved swiftly to reinforce the resolution, closing Faxa Bay, an important fishing area, for conservation reasons. Iceland's decision expressly linked conservation with enclosure and protection from the distant-water fleets. The British fishing industry responded by banning Icelandic trawlers from landing in their ports, a boycott upheld by the unions that unloaded the boats. The British industry overlooked that the Faxa ban also applied to Icelandic trawlers, not just to the foreign boats.

Iceland was not the only country unhappy about British boats fishing in its waters. Norway had tried for decades to restrict British fishing. Finally, in 1935, Norway issued a decree expanding its waters to a four-mile belt seaward from straight baselines drawn between 48 points on Norwegian promontories, islands, and rocks. The result was the enclosure of a large area of water that had formerly been regarded as the high seas and an important British fishing ground. Britain applied to the International Court of Justice at The Hague for a ruling and in 1951 the court upheld the Norwegian position.[3] The decision was one of the first successful attempts by a country to restrict foreign fishing on territorial grounds and it was a substantial blow to all nations with distant-water fishing fleets, including the U.S.

The International Law Commission, which had been studying the territorial and fishing questions since the late 1920s, was delighted to have the issues resurrected. It issued a draft report in the summer of 1952, recommending the creation of an international framework, under the Food and Agriculture Organization (FAO) of the UN, to come up with regulations to protect fish resources from waste or extermination on the high seas. The new group's regulations would be binding. It also recommended that territorial limits be expanded to six miles. The draft recognized that the only way to protect local fish from the foreign boats was to enclose their waters.

It was just the encouragement Ecuador, Peru, and Chile needed to act against what they viewed as illegal fishing and whaling in their waters. At a series of meetings, starting in Santiago in 1952, they began discussing a 200-mile regional zone, with their own regional fisheries research organization and the right to control passage through their waters. They were moving decisively in the direction of enclosure.[4] Chile was unhappy about illegal whaling ships in its waters, outside the territorial boundaries and seasons set by the International Whaling Commission.[5] Peru and Ecuador were concerned about American tuna boats and seized any they found fishing within their waters.[6] Tensions increased as more tuna boats were arrested and fined.

The tuna industry pressed the State Department for a solution. The Departments of State, Defense, and the Budget were all opposed to any tariff relief for fishermen. However, the U.S. could not allow three Latin American countries to create a zone that would prohibit fishing boats, perhaps setting a precedent for trying to ban military and merchant marine vessels. International law could not be made on a regional basis. With the Latin Americans moving so quickly, the U.S. decided, "out of desperation" as the State Department later put it, to host its own meeting.[7]

They asked the FAO to host the meeting at the organization's new offices in Rome. They would attempt to outflank the Latin Americans, by moving the issue to a European forum where the American position would receive more support from other distant-water nations.[8] Once that was accomplished, the U.S. moved forward with shaping an agenda that would achieve its objectives.

The International Technical Conference on the Living Resources of the Sea began April 18, 1955, at the FAO headquarters in Rome. Delegates from 45 countries attended, including the Soviet Union, as well as scientists from regional and international organizations. The emphasis of the meeting would be "technical," to provide expert advice on fisheries to the International Law Commission (ILC), which would be meeting in Geneva six weeks after the Rome meeting. "It was emphasized that the Conference should not discuss matters of a legal or political nature," according to the introduction of the published scientific papers.[9]

Most of the nations in Rome, including Japan, the Soviets, and Spain, all sided with the U.S. in not wanting to grant coastal nations any special interest in the resources of their shores. To the surprise of the Americans, the Soviets, after some differences on organizational matters, voted consistently with the U.S. and Britain.[10] After all, with the enormous fleet they were building, they were interested in upholding the freedom of seas along with the imperialist U.S.

The meeting preempted any action on creating an international organization, preventing Peru, Ecuador, and Chile from setting up a regional zone with its own law. The Latin Americans did win an important battle, though: recognition of the right of coastal states to regulate their waters, potentially jeopardizing the rights of distant-water fishing nations to continue to fish. It was a Pyrrhic victory, because events were moving quickly. Overfishing had to be documented scientifically before fishing could be curtailed. Britain's chief fisheries scientist, Michael Graham, had wanted fishing restricted as fisheries developed so scientists could study the impact of the gear on the stocks. But the race was on to develop fisheries, deploy them, and lay claim to the fish in the ocean. Territorial claims needed to be established to prohibit other boats from entering new fisheries.

The distant-water fisheries were the big winners. The American tuna fleet, as well as the distant-water fleets of Britain, Japan, and a host of other developed countries, could continue to fish off the coasts of poorer, less-developed countries. Foreign fishing could only be halted when scientific studies proved that overfishing was occurring—the most significant decision in the history of natural resource management. Given the vestigial state of fisheries science in 1954, with most states, provinces, and countries just starting to build natural resource management institutions, it was an impossible standard to meet.

Iceland was disappointed that the issue of territorial waters had been deflected. The British were boycotting Icelandic fish, but the Soviet Union was buying more of the catch. By 1955, it displaced Britain as the largest single importer of Icelandic fish and became Iceland's second-largest trading partner. To the frustration of the Americans, some in Iceland argued that this expanded trade had relaxed East-West tensions and it was time for the U.S. troops to go home (which would happen in 1958).

Iceland elected a leftist government in 1956. Its members wanted to abrogate the bilateral treaty of defense that had signed with the U.S. just five years earlier. The new government made it clear that it would not jeopardize Iceland's membership in NATO, but that it wanted American troops to leave. It was an unpleasant election for the Americans. They were now partners with a government that included pro–Soviet Socialists. "At US bidding, NATO headquarters at Paris immediately stopped circulating confidential documents to Iceland, on the grounds that they might end up in Socialist hands," wrote historian Valur Ingimundarson. The fishing fight not only challenged the political cohesion of NATO and Western military plans in the North Atlantic, it also threatened to hand the Soviet Union a propaganda victory in the Cold War. "Ultimately at stake was the continuation or reversal of Iceland's political, military, and economic integration into the Western alliance," concluded Ingimundarson.[11]

In the meantime, the international regulatory process that had been set in motion in 1949 now entered a formal new phase. The Eleventh General Assembly of the United Nations passed a resolution on February 21, 1957, to hold a conference that would look at legal, technical, and political aspects of the sea. The first Law of the Sea Conference opened March 11, 1958, in Geneva. The countries that participated claimed territorial seas ranging from three to 200 miles, with an assortment of those who didn't want any limits at all.[12]

The meeting had been forced by territorial questions over fishing. The coastal states wanted to establish ownership of the fish off their shores, which the distant-water nations opposed. There was a vote on a compromise of setting a 12-mile limit but it failed to get a majority of votes. There was also a new disagreement, over ownership of the mineral resources of the deep seabed. Scattered all over the seafloor were manganese nodules containing nickel. The sea also contained dissolved minerals such as gold and silver. A farmer could till the top few inches of land, but the whole water column was a potential source of riches! It would just be a matter of time before the wealth of the seas could be mined economically.

"This stimulated two very different agendas," writes economist Rögnvaldur Hannesson. "First, the pressure from coastal states wanting to establish ownership of the fish resources off their shores continued to intensify in tandem with the advances in fishing technology and increasing pressure on fish stocks. Second, opposition arose against the possibility that certain states or private companies would get hold of the mineral resources of the deep seabed."[13]

Wilbert Chapman was an advisor to the 1958 meeting (as he had been in Rome in 1955). It was clear to him that fisheries, which had been in the forefront of the battle over territorial waters, were now in the backseat, overshadowed by military developments in the Middle East. The Departments of State and Commerce, as well as the White House, had written off the American fishing industry. They wanted to support foreign fisheries, not the domestic industry. "Support of a domestic industry by the United States Government either by tariff or subsidy would have the equal effect of lessoning the advantage that the foreign industry will have in this market," Chapman wrote after the meeting. "To hell with the fisheries."[14]

Iceland's delegates had been hopeful as they traveled to Geneva. But the meeting ended with no agreement and Iceland was angry with the U.S. for not supporting their attempt to expand their seas. The government returned home and, once again, unilaterally expanded its territorial sea to 12 miles. Britain not only refused to recognize the law, it sent warships to protect its fishermen in the 12-mile zone.[15] The actions resulted in the first so-called cod war that lasted until February 1961, and NATO was in the thick of it.

Iceland demanded that Britain recognize its territorial waters. It even threatened to withdraw from NATO. This horrified the Americans, who were in Iceland to assist the country in its defense; obviously the U.S. would not get involved in a conflict with Britain's navy to defend Icelandic fishing limits, which it did not officially recognize. Iceland and Great Britain were both among the original charter members that signed the North Atlantic treaty. The peculiar result of the situation was that Iceland, a member of NATO, in reacting to a trade boycott from another NATO country, developed strong trading ties with the Soviet bloc. When it comes to fish, there is always another agenda.

The involvement of the Soviets, the Americans, the British, and NATO emphasize the complexity of international fishery disputes, according to geographer Bruce Mitchell. "In this situation the specific resource question becomes more of a pawn in a complicated match that includes political, economic, and cultural objectives."[16] Ingimundarson agrees, arguing that both NATO and the U.S. wanted to maintain Iceland's strategic tilt to the West.

France, the Netherlands, and West Germany all had vessels fishing on the continental shelf adjacent to Iceland. These countries protested the extension but advised their fishermen to respect the decree. Great Britain took a different stance. Declaring that no legal justification existed for such unilateral action, it used the Royal Navy to protect British trawlers from Icelandic Coast Guard vessels.

Following the 12-mile boundary, Iceland introduced a 50-mile limit. Even that was only a temporary measure by the Icelandic government until it could extend the limit to 200 miles, which it did in 1975 provoking the third and last modern cod war. An agreement was finally reached to reduce the number of British boats in Icelandic waters and to reduce the British catch; agreements were also necessary with Norway and Belgium. Icelanders were eventually successful in establishing the 200-mile boundary as an exclusive economic zone (EEZ). The British fleet left Icelandic waters in 1976, followed by the West German fleet the next year. Belgian, Faroese, and Norwegian boats were allowed to continue fishing in limited quantities.[17] It is hard to detach the claws of historical fishing regimes. Still, many distant-water trawlers were laid up in British and German ports, unable to fish.[18]

The fishing industry had new markets but Iceland was still not self-sufficient. Nearly a third of the revenue from exports went to producers in the form of subsidies in 1957 and 1958, compensation for the unrealistically high official exchange rate. Iceland received $20 million from the U.S. in grants, $15 million in loans, and $73 million in expenditures to the air base. It had also negotiated loans with European countries and the International Bank for Reconstruction and Development. Foreign debt increased sharply and loan payments accounted for 10 percent of the export earnings.[19] There was enormous pressure to do something to relieve the situation.

"Iceland was extremely skillful at extracting economic aid from the United States, receiving over $70 million in grants and loans between 1948 and 1960," wrote Ingimundarson. The air base pumped up to $15 million a year into the economy, about a fifth of Iceland's foreign currency earning. Economic aid increased the national per capita income about 11 percent between 1952 and 1957. "This was the price the United States and NATO paid for political and economic stability, for their military presence, and for limiting the influence in Iceland of the Soviet Union and its supporter, the Socialist Party."[20]

Iceland received $38 million in Marshall Aid between 1948 and 1953, in relative terms more than any other European country. And after a hiatus lasting two years, the United States granted Iceland an additional $34 million between 1956 and 1960. This aid served the same purpose as the Marshall Plan: to ensure economic and political stability and head off communist encroachment. In addition, the Americans gave Iceland substantial loan guarantees to buttress its currency reserves, enabling the Reconstruction Government to enact its market reforms.

The northern herring stocks finally recovered and catches were strong, allowing the fleet to diversify. They were able to adopt the power block and nylon nets, elements of the global transformation in fishing: the expansion of purse seining. But purse seines allowed whole schools of fish to be caught, leading to a population crash by the late 1960s.[21]

Subsidies continued to play a part in Icelandic fisheries management through the 1970s, although they have been substantially diminished, according to an analysis by scientist H. P. Valtýsson. The remaining subsidies are in the form of personal income tax breaks. Fishermen can withhold a certain amount from taxes for each day registered at sea. The tax break costs the government about U.S. $23 million annually in lost revenues. Indirect government expenditures related to the fishing industry have been estimated to be approximately U.S. $38 million. This includes fishery related education, various monitoring and research institutions related to the fishing industry (including marketing), the Ministry of Fisheries, and part of the operations for both the Ministry of Environment and Ministry of Foreign Affairs. Expenditures directly aimed at managing the marine stocks—including the Marine Research Institute, Coast Guard, Ministry of Fisheries, and Directorate of Fisheries—have been estimated at around U.S. $16 million per annum. In turn, the Icelandic government receives about U.S. $8.5 million per annum from the fisheries from various fees and licenses.[22] The subsidies were mostly financed by imposing duties on imports. Luxuries were taxed very heavily—in some instances up to 200 percent of the value—while articles that were important in the cost of living index were in some cases totally duty-free.[23]

THE AMERICANS

As the 1960s wore on more Soviet boats, as well as the boats of other nations, appeared in American waters. Something needed to be done, but there was no consensus on what. The Departments of State, Budget, and Defense continued to insist on the freedom of the seas for all boats, including Soviet fishing boats. It was not a popular position. Congress was confronted with dozens of bills, many of them recommending increased funding for marine research. A bill from Senator Claiborne Pell of Rhode Island (sponsor of the Pell Grant) finally passed, establishing Sea Grant colleges and programs, to initiate and support programs of education, training, and research in the marine sciences, and offer advisory services.

Sea Grant had four goals: first, to acquire the ability to predict and ultimately control phenomena affecting the safety and economy of seagoing activities; second, to fully exploit the resources "represented by, in and under the sea"; third, to utilize the sea to enhance national security; and fourth, to pursue scientific investigations to understand marine processes and resources. Much of the initial research would focus on ways to catch fish more efficiently, find new sources to exploit, increase survival of fish in hatcheries, and find ways to utilize the waste that came from fish plants. With this new source of government money, the function of educating fishermen moved from the industry itself into the universities and research offices. As the number of fishermen shrank, the number of researchers, assistants, bureaucrats, and support staff would continue to grow.

For the universities, the bills meant a rich flow of federal funds for a variety of oceanic projects. Universities scrambled to put researchers from different departments into new programs to vie for the new funding. This was so challenging that only three universities were able to put together an adequate plan for the first round of funding.[24] Knowledge of the sea and fish stocks was fragmentary; there was no integration of information. There was very little understanding of the ocean system, and certainly no awareness of where the limits of fishing would be—if indeed, there were limits. After all, technology might help humans evade limits.

The chorus of complaints about the Soviet boats off both American coasts grew louder. The State Department continued to resist the weakening of the doctrine of the Freedom of the Seas. The Bureau of Commercial Fisheries also resisted, arguing that expanding the territorial sea to 200 miles would mean areas of the ocean would not be fished for "maximum utilization of marine resources."[25] Fish that were not caught had no value.

The New England fishermen fumed helplessly while Soviet boats scooped up record numbers of haddock, then moved south to target Atlantic mackerel and herring. Total haddock landings had declined steadily from a record high of 339 million pounds in 1936 to 151 million in 1954.[26] The U.S. industry was equipped with smaller, older boats, self-financed, and with limited equipment and technology. Recent evidence suggests that catches had been on a downward trend since 1860.[27] "The programs to increase demand for groundfish and to decrease the costs of fishing had few effects on the groundfish industry because they were incorrectly conceived, badly implemented, and too small to copy with very large problems," argued Margaret E. Dewar in her 1983 analysis of New England fisheries.[28] People were eating more fish sticks, but those could be made with fish caught anywhere in the world—from Canada, Iceland, or Norway. The market for fish had increased, but New England did not have the fish to supply it at a price that buyers would pay.

President Johnson appointed a commission on marine sciences, engineering, and resources in January of 1967, under the chairmanship of Julius A. Stratton. The commission documented a litany of fishery problems, including state and federal agencies competing with each other and no long-term rational goal for ocean development. It called for the creation of a "wet NASA," which resulted in the creation of the National Oceanic and Atmospheric Administration (NOAA) in 1970. NOAA was supposed to provide a central and integrated focus for conserving and managing the nation's coastal resources. But as historian of science Sara Tjossem points out in her analysis of postwar oceanographic science, there were still many agencies with a finger in the pie, or a thumb on the scale.[29] Nine new acts were passed, giving sweeping powers to federal agencies, including the Coastal Zone Management Act and the Marine Mammal Protection Act, and, eventually the Fisheries Conservation and Management Act in 1976.

There was no shortage of blame for the dire state of American fishermen. Wenk cites a long list of problems, the most important of which was a low rate of return on money invested into fishing. Anybody could buy a boat and go fishing; there were too many boats, even though many fishermen were making little money. Since there were no restrictions on who could enter a fishery, the government created inefficiencies to slow the catch. The inefficiencies increased costs, handicapping the introduction of technological innovations. Local and state regulation was fragmented and confusing.[30] The state waters ended at three miles. There was no attempt at management beyond three miles.

The industry held a large conference in Seattle in 1968 at the University of Washington, dedicated to identifying the causes of stagnation. Senator Magnuson gave the keynote talk and close to 300 industry officials, government employees, and university researchers attended. The conference concluded that Americans were eating more seafood, but that fishermen were unable to supply the domestic market, forcing the processing and distribution sectors to depend on foreign fish. Fish prices were held down by the price of other commodities, like pork and chicken, while production costs were increasing. "This situation exists even though the United States has a large stock of fish near its shores, has the ability to operate any kind of fishing equipment anywhere in the world, and has the largest market in the world for fish products," the report concluded.[31]

It was not in the best interest of the nation to allow the industry to continue to deteriorate and be dependent on foreign sources of fish. The industry looked to the federal government to create more positive policies to help it expand, to help the fleet get more efficient, find markets for new fish species, and to reorganize management at the federal level, eliminating fragmented state and regional jurisdictions.[32] Another problem was a federal law requiring that boats fishing in American fisheries be built in the United States, where shipbuilding costs were higher than in many other nations.

In attendance at the meeting was Wilbert M. Chapman, now working for Ralston Purina Company in San Diego as their director of marine resources. He blamed the difficulty the industry faced in expanding on policies of the Departments of State and Defense. The BCF budget was constrained by pet projects from lawmakers. "Management" meant introducing inefficiencies that made it costlier to fish.[33] He defended the industry, saying it was not "inefficient, technologically retarded, economically incompetent . . . [or using] the hunting technology of our Neolithic forebearers." Instead of gloom and doom, he said the volume of fish production was increasing at 6 to 8 percent a year, and that development of unutilized or underutilized species would allow ocean fish production to increase "on a sustainable basis by a factor of somewhere between 4 and 40 times."[34] But Chapman's statistics reflected the global growth of fishing and masked the stagnation of the American industry.

The industry was beset by problems. But something else was pointed out at the Seattle meeting by Hiroshi Kasahara, a Japanese-born scientist working for the International Pacific Salmon Fisheries Commission created in 1953 by Canada, Japan, and the U.S. Kasahara has written widely and perceptively about fishery development in the North Pacific. Both the U.S. and Canada were more interested in protecting their coastal fisheries than in getting into distant-water fisheries for fish nobody had heard of. Canadian and U.S. fishermen were interested in high-value species: salmon, halibut, herring, and crab. It was the Koreans, the Japanese, and the Soviets who were expanding in the North Pacific, catching fish that did excite American consumers.[35] Canada and the U.S. did not have boats capable of catching those species. After a decade of battling and losing over tariffs, the industry's political focus was to reduce salmon and halibut bycatch in the Japanese and Soviet pollock catches, not necessarily to catch pollock themselves.

Kasahara estimated the Japanese catch in the North Pacific at roughly 8.5 million metric tons by 1971, while the U.S.S.R. took a little over 2 million metric tons. His figure for the South Korean catch was 830,000 metric tons, with no estimates available for the Chinese catch from the East China Sea and the Yellow Sea.[36] The Canadians and the Americans, despite their proximity to the fishing grounds, were small producers, less than 1 million metric tons each.

How many fish could be caught? Nobody knew, but the assumptions were that the sustained harvest might be in the neighborhood of 200 million metric tons. Some thought that estimate too conservative, such as Chapman, who suggested 400 million metric tons. The rosy projections about the potential harvest from the seas were part of postwar optimism about the benefits that science and technology were bringing to humankind. Fish could be found, they could be caught, they could be preserved, and they could be marketed. Increasing the fish catch would be easy. Finding tuna in the central Pacific would happen quickly. This was the period when the President's Science Council assumed that underwater nuclear power plants would help provide energy in the future, when it seemed that scientists not only understood the natural world, but were ready to take command of it. And for a while, it seemed as if nations could catch as many fish as they wanted.

This connection between the development of fisheries science and the creation of industrial fishing has been lost to sight, as historian of science Jennifer Hubbard has pointed out. For most of the century, fishery scientists were partners with fishermen, all of them seeking to increase the catch. As Hubbard notes, "many people—even fisheries biologists themselves—believe that fisheries biology began and continued as a conservation-oriented science seeking to identify and remediate overfishing."[37] In fact, postwar fisheries science focused on helping fishermen find fish more efficiently, process it into new product forms, and sell it to the public. As fish became scarce, the relationship soured and became more adversarial, especially in New England.

The real turning point in the battle over foreign fishing came at the 1974 meeting of the Third United Nations Conference on the Law of the Sea (UNCLOS III) in Caracas. There was a widespread realization that overfishing and depletion were occurring, with no effective international organizations to manage, wrote legal scholar William Black.[38] Developing coastal nations were frustrated that they could not develop their own fisheries, in the face of widespread depletion by foreign fleets—be they Japanese, Soviet, American, or European. A year earlier, during the Arab oil embargo, the price of fuel had skyrocketed, directly challenging profits for the factory processing ships.

Support was building in many countries for a 200-mile limit that would eliminate foreign fishing. The American tuna industry, the only fishery that had expanded into the high seas, was increasingly isolated politically. Salmon and groundfish fishermen on all the coasts, as well as sport fishermen and community residents, wanted the foreign boats gone before they caught all the fish. The Departments of State and Defense held out as long as they could, but the pressure grew and could not be resisted.

With no clear idea of how many fish might be available to catch, Congress created a series of subsidies. The 1970 Fishing Vessel Capital Construction Fund Program allowed tax deferments that let fishermen set aside money for future vessel upgrades or new construction. The Fishing Vessel Obligation Guarantee Program passed in 1973, guaranteeing loans for up to 87.5 percent of the cost of reconditioning an existing vessel or constructing a new one. "The government was promoting a bright future for the Pacific groundfish industry, creating a bonanza for fishermen, bankers, and shipyards along the entire coast," wrote scientist Donald Gunderson. "Yet this would eventually turn out to be just one more unsustainable boom based on false assumptions."[39]

The Fisheries Conservation and Management Act—now known as the Magnuson Stevens Act, or MSA, after the two senators who did most to shape it, Warren Magnuson and Ted Stevens—passed in 1976. Author David Helvarg argues the Act was not about conservation of fish so much as it was an assertion of exclusive U.S. fishing rights on the continental shelf.[40] There were environmental groups pushing for new legislation, but they were more interested in policies around ocean dumping, marine mammals, and water and vessel pollution, not fish, according to James P. Walsh, a one-time aide to Magnuson who participated in the negotiations.[41]

The Act created a vacuum, which fishermen and fisheries developers rushed to fill. "U.S. fleets grew and modernized at a breakneck speed," wrote historian Michael Weber in his account of this period. "Private bankers, investors attracted by tax shelters, the [National Marine Fisheries Service (NMFS)], and other federal and state agencies provided financing for the largest buildup of U.S. fishing vessels in the country's history."[42]

MSY was further solidified with the passage of the Act in 1976, when it was declared to be the "best available science." This is embedded in the seven national standards, and was first drafted on the back of a large manila envelope, according to the man who probably wielded the pencil: Dayton Lee Alverson, who was now a senior administrator at the Bureau of Commercial Fisheries during the 1960s.[43] In his memoir, he recalled that a group of scientists and policy makers met to discuss how the new system of federal councils was going to work. They were worried that politics might trump fish conservation, so somebody came up with the idea of using "the best available science."[44] This would be a variation on MSY, optimum yield (OY). The first standard was to prevent overfishing and ensure optimum yield from each fishery, and to be based on the best available science. A

Fig. 7.1. Dayton Lee Alverson helped define what came to be called the "best available science" in the 1976 Fisheries Conservation and Management Act.

problem, of course, was that the best available science at the time was not very good, as oceanographer Alan Longhurst has written.[45] Scientists were soon suggesting that MSY was interpreted too rigidly, with too much of a focus on harvest, and not enough on conservation.[46]

The national standards were kept loose intentionally, to allow each of the eight new councils flexibility in dealing with their fisheries, according to Weber. His account is supported by Christopher M. Weld, the attorney who wrote the original act. In a 1977 paper, Weld said there was great confusion around MSY.

"It is highly doubtful that either Congressional staff personnel responsible for drafting the definition or the Committee members who approved it had more than a tenuous grasp of the concept," Weld wrote. "More than once in Committee session the sentiment was expressed 'Well, you've got to start somewhere,' as if MSY were a fixed reference point. Unfortunately, it is not."[47]

THE PACIFIC ISLANDS

The decision to allow Japanese-caught tuna to be delivered to American canneries in American Samoa was an important key to the reestablishment of Japan as one of the world's leading fishing nations. The islands were to be modernized and the residents trained to participate in a capitalist economy. It was not economic for American boats to deliver tuna to the islands, at least during the 1950s and 1960s, and the fish plant needed a reliable supply of fish—most consistently from Japanese and Korean boats. The isolation that made the islands so attractive to the U.S. continued to make it impossible for the islanders to escape their colonial past or to control the exploitation of the tuna stocks in their waters. A second decisive American action was the decision to declare tuna stocks "highly migratory," ensuring that both Japan and the U.S. would continue to dominate the tuna catch in the region.

The fishery in the island waters is one of the world's largest, in terms of both catch and value.[48] More than 2 million metric tons of tuna are taken each year, about half of the global tuna catch, with an annual value of $3 billion in U.S. dollars (as of 2005). Yet little of the wealth in the fishery is seen in the islands. The island governments built boats and tried to compete in the fishery. But just having the fish in your backyard is no guarantee that you can catch them and sell them in the competitive market.

The global tuna catch would escalate sharply with the widespread introduction of new synthetic nets, lighter and stronger than natural fabric nets. Hauling the nets in by hand was slow and laborious and it limited how many sets could be made in a trip. San Diego tuna man Mario Puretic solved the problem in 1955 by creating a hydraulic power block to pull in the net. The big new nets could handle large volumes of skipjack and yellowfin.[49] The technology was rapidly adopted in Norway, Iceland, Spain, and Portugal. The gear could be adapted to traditional boats, increasing the catch capacity and reducing the size of the crew.[50] One big advantage was that boats did not need bait, since catching local bait was both expensive and politically sensitive. It was also becoming increasingly obvious as boats explored more of the Pacific that bait was hard to find, perhaps explaining why tuna evolved to go long periods of time without eating then moved into a feeding frenzy when bait was found.

The U.S. may have been forced to give up on the concept of the Freedom of the Seas, but it took steps to retain its territorial claim to high-seas tuna. The MSA included a provision that defined tuna as a "highly migratory" species, a move designed to make the United States part of any tuna regulatory regime in the Western Pacific. The U.S. argued that it acted to conserve tuna resources, but this ignores the long history of American attempts to expand its tuna fisheries, and its self-interest in maintaining a claim on the high-seas tuna stocks.

It took until 1982 for the UNCLOS III to be signed, establishing EEZs of 200 miles and giving coastal nations control of large areas of the ocean. This was important for many nations, but especially for Micronesia. A 200-mile zone would establish their claim to one of the great ocean resources, high-seas tunas. They could join other modern nations in the growing international fish trade, or they could charge "rent" for fish caught in their waters by foreign boats.[51]

The islanders started to take the steps to declare a 200-mile EEZ. No matter how it was plotted on a map, a 200-mile zone would exclude both the U.S. and Japan, the two countries harvesting the bulk of the tuna. In a move that Wilbert Chapman (who died in 1970) would have approved, the U.S. argued that tuna were "highly migratory" and therefore better regulated by an international commission composed of fishing states.

This approach ensured that the United States, as well as Japan, would be part of the management process. A fishery management scheme composed of members of the EEZ, by contrast, would have excluded the big powers.[52] As historian Kimic Hara argues, the post–Cold War order in the Pacific reflects American desire to have unfettered access to the world's oceans.[53] This is certainly true of tuna. Both the United States and Japan have strategic reasons for insisting on access to the fish stocks in the region. By insisting on access for themselves, they opened the door to other nations to expand into the fishery, especially Taiwan and South Korea.

American boats began entering the Western Pacific fishery in the 1970s, driven by declining catch rates and deteriorating relations with Latin American countries. Despite the growing number of American boats fishing the island waters, the United States refused to recognize the claims of expanded jurisdiction until 1988, and a convention to regulate the catch was not signed until 1995.[54] One of the factors that prompted the United States to finally capitulate was that some island countries were starting to establish access agreements with the Soviets. The U.S. agreed to a five-year treaty and to pay approximately 9 percent of the value of the catch as an access fee.[55]

Low wages attracted companies to American Samoa, but soon wages were even lower in cannery jobs in other countries. Writing in 1985, two researchers reported that labor was 40 cents an hour in Thailand and $3 an hour in Samoa. There were additional advantages in Thailand: long-term tax concessions, limited responsibility for employee health and welfare, and relatively weak environmental restrictions on cannery emissions. It was a similar story in the Philippine tuna industry, which got its start through U.S. A.I.D., financed by U.S. tax dollars.

It was not just the Soviets, Japanese, and Americans who sold their industrialized fishing technology. German experts traveled to Thailand in 1959, according to historian Franziska Torma. With fishing becoming more competitive in the North Atlantic, the Germans wanted to use Thailand as a bridge to launch fisheries in the Indian and Southeast Asian seas. "In the 1970s, however, the growing fleet of Thai-German trawlers had almost emptied the Gulf of Thailand and the adjacent seas," Torma wrote. "Overfishing and ecological degradation were major, yet unintended consequences of this project."[56]

Throughout the 1970s and 1980s, Thailand emerged as a leading exporter of canned tuna. "They wasted nothing: tuna blood and flesh regarded as unsuitable for canning [were] made into pet food, the water in which the tuna was cooked was used to make a concentrated soup, and bones were ground in fertilizer," wrote John Butcher in his account of the impact of Japanese fishing in Southeast Asian waters. "And they paid low wages to their workers, nearly all of whom were young women hired on a daily basis. Since canning tuna is a highly labor-intensive process—skinning the tuna and cutting the loin meat from the fish must be done by hand—this gave Thai canneries a great advantage over many of their competitors."[57]

This decision to exempt companies from paying American-scale wages in fish plants paved the way for other U.S. companies to move to the islands, setting up manufacturing companies that could ship clothing to the U.S. without tariffs. These conditions allowed the establishment of sweatshops that have a long history of abuse of workers.[58]

Since the early 1990s, all of the Pacific Island countries have tried to increase domestication of the tuna industry and almost all of the efforts have failed. The islands tried to invest in harvesting and canning, but those are two of the least profitable segments of the highly competitive tuna industry. The money is in distribution and marketing, which are controlled by Japan, Korea, China, Taiwan, and the U.S.[59] As biologist Gary Sharp has noted, the complexity of international negotiations, and the ability of fishing nations to extend the negotiations process, "allowed the emergence of a new generation of international corporations with unusually complex marketing systems. These evolve continuously in efforts to cope with changes in legal ownership, access, and regional and local responsibility for resource management."[60]

The success of the initial canneries in American Samoa led seafood companies to continue to ship jobs overseas. By 1985, only one cannery was still operating in the U.S. The rest moved to American Samoa or Puerto Rico, as well as other low-wage countries. Relocation of U.S. tuna canning operations meant the loss of an industry that produced $1.5 billion in food products.[61] Among the closures was the Elmore Cannery in Astoria in 1980, built by the Columbia River Packers Association. The plant had been the largest employer in Clatsop County.[62] It was the end of an era.

Even the cheap labor in Samoa could be undercut. As Thailand industrialized, it offered tax concessions, limited responsibility for employee health and welfare, and weak environmental regulation of cannery emissions. "The processing of tuna at all major offshore locations is subsidized in part by

U.S. taxpayers with an even larger cost paid by coastal U.S. tuna fishermen who through their tax payments have helped subsidize the relocation of their only market," two California researchers reported in 1985.[63]

Negotiations over sustainable management for tuna between the distant fishing nations (especially the U.S., Japan, Taiwan, and South Korea) and the island governments did not begin until 1995. The framework for the Western Central Pacific Fisheries Convention was adopted in 2000. The objective of the convention was to ensure, through effective management, the long-term conservation and sustainable use of fish stocks in the region.[64] The island states themselves built boats during the 1980s, hoping to capitalize on their proximity to the fish and create local jobs. They entered the fishery at a time when the industry was extremely competitive and the harvesting sector was being squeezed by low prices.[65]

A brief review of the development of fisheries in the Western and Central Pacific Fishery Commission reveals a number of efforts to control the rate of fishing and the number of boats, at the same time as they try to achieve consensus on how to implement a 25 percent reduction in the catch of bigeye and a 10 percent reduction in yellowfin. Economic studies showed that fishing effort was substantially above optimal levels, making the fishery less profitable than it could have been.[66] More troublesome, growing use of highly efficient purse-seine nets substantially changed the age structure of the stocks, resulting in a lowering of the MSY estimate for yellowfin tuna to 4 million metric tons, down from 4.5 million metric tons.[67] Implementation of the fisheries convention was slowed by competing interpretations of the text and uncertainty about the extent to which the islanders would be allowed to increase their purse-seine fishery. Residents are disappointed that they have not commanded a greater share of tuna resources.[68]

The failure in the Western and Central Pacific is the result of the overall shape of modern fisheries management, established in Rome in 1955, on a supposed scientific foundation that had been adopted through political maneuvering and driven by Cold War concerns. Fishing could not be restricted until there were signs that stocks were overfished; then regulations could be implemented to slow the catch. Despite the voluminous criticism of the failures of MSY, the pattern established at Rome—that scientific evidence of overfishing was required to restrict fishing—continues to be embedded in the legal framework for international fishery commissions, making it more difficult to create fisheries that will operate in a sustainable manner by limiting the number of boats.

CHAPTER EIGHT

Enclosure

Not only is the past recalled in what we see; it is incarnate in what we create.
—David Lowenthal[1]

Throughout the 1960s, nations created conflicting territorial claims hoping to restrict foreign fishing and whaling in their waters. Japan, the U.S., and Great Britain upheld the historic claim of three miles. India claimed six miles, Mexico claimed nine, while the U.S.S.R., Venezuela, Panama, Canada, and the United Arab Republics claimed 12 miles. Iceland claimed 50 miles, but was threatening to expand to 200. Off Central and South America, Chile, Ecuador, Peru, Costa Rica, and El Salvador claimed up to 200 miles of exclusive fishery jurisdiction.[2]

It all changed, almost overnight, as nations adopted new 200-mile zones. An estimated 99 percent of the world's commercial fishery stocks were within the new zones. Distant-water nations either had to find new and productive fishing grounds outside coastal state jurisdiction, or they had to negotiate to gain access to the resources of coastal states. Politicians, economists, and biologists all backed the new regime, seeing it as a first step to tackling the mess of the competing claims. "With proper training and assistance, developing nations would be able to identify fishery management objectives that suited their national goals, whether economic, social or biological," wrote legal scholar William Black.[3] That sounded good, but the management objectives would all revolve around harvest and the development of industrial fisheries. Not to catch the available fish was to waste them. Modernization meant industrialization, as quickly as possible.

It is during the period from 1945 to 1958 that many of the international institutions to manage fisheries and whaling were framed. Most international commissions were bilateral and multilateral organizations, with very limited authority. The states controlled how many boats would be allowed to fish. Efforts to limit licenses and boats were slow to emerge, especially in the U.S. where fishermen fought limited entry well into the 1980s.

The postwar development of deep-sea fisheries was a frenzied free-for-all, as governments subsidized the creation of an industrialized fisheries system, driven to catch as many fish as fast as possible to sell for as little as possible. Despite all the rhetoric about how expanding the protein taken from the sea would feed the world's poor, much of the fish (especially Peruvian anchovy) was turned into fish meal to feed the growing numbers of chickens, pigs, and cows in Europe and North America.

Fishing played very different roles in different national economies. It was of central importance for Iceland and Japan. It played a very small role in American economic life, but it was vital to the region of New England, where it had a deep historical role in the founding of the states.[4] Japan, the Soviets, and the South Koreans were interested in distant-water fisheries for low-value species like pollock. But Canada and the U.S. wanted to protect their high-value coastal fisheries for salmon and halibut. Japan and the U.S. both wanted to maintain access to high-seas tuna. It was a difficult set of interests to reconcile, and the contradictions are reflected in most fishery management programs.

Nations began expanding their exclusive economic zones, most to 200 miles, setting off a scramble to build fishing fleets. The money was a massive jolt to the always dynamic and volatile world of fishing. Government money was followed by private money. Shipyards across the world started turning out fishing boats. Hundreds of thousands of people went fishing, at the same time as unprecedented technological capacity was being created. Fishermen would soon be able to find and catch fish anywhere in the oceans. Information about technology transferred with great rapidity, publicized by government scientists, often reporting on the discoveries their research vessels were finding and the work done in their laboratories.

Between 1950 and 1975, the world economy grew almost 5 percent per year, and 3 percent per year per capita.[5] Fishing grew 8 percent a year through most of the 1950s, which was interpreted as evidence that stocks were being well managed and scientists were on the right track in their work.[6] Nations went fishing in confidence that they were modernizing and industrializing, and secure in the knowledge that the massive catches were "surplus allowable catch" to what fish needed to sustain their stocks.

The U.S., Japan, Iceland, and the Soviet Union used fisheries to further their political objectives as the world increasingly divided into two Cold War camps. For the United States, a cornerstone of postwar foreign policy was preservation of open seas for American fishing boats, a proxy for military vessels. Thanks to the 1955 meeting in Rome, distant-water boats would continue to fish, even if they had to pay access fees. It was also important that both Japan and Iceland continued to tilt toward the West; it was worth the loss of fishing jobs in California and New England. For Japan, access to the high seas was vital to supplying the nation with food and also a key component in the recovery of the national economy. The Soviets desperately needed protein and the industrialization of fishing, taking it to a scale unimagined, was a propaganda coup over the Americans. Iceland had long wanted to limit foreign boats fishing in its waters, securing its resources for its own fishermen. Fishing was never just about fish.

The U.S., Japan, and the Soviet Union all scrambled to set up an extensive network of foreign alliances around fishing. The Food and Agriculture Organization (FAO) of the United Nations was also sending Western "experts" to third-world countries, bringing the Western ideology of progress. These scientist-ambassadors, as oceanographer Alan Longhurst calls them, promoted Western ideas about science and scientific practices in agriculture, oceanography, and fisheries biology.[7] The visiting experts helped developing countries build large-scale, mechanized fisheries.

With the collapse of California sardines, the canning industry was dismantled and moved to the next crop of fish waiting to be harvested. Peru wanted to develop its vast anchoveta (*Engraulis ringens*) resource into a fishmeal industry (it also wanted to build its own whaling fleet) and companies bought equipment from California. Equipment also went to South Africa, turning a country with a negligible fish pack into one of the major fish canning countries of the world. By 1957, South Africa was producing some 2.5 million cases of pilchards. The industry moved into the markets formerly supplied by California sardines.[8] The boats were mobile, the processing equipment could be moved, and the first company to invest and begin to fish could make the most money.

There were ominous signs that the world's oceans were being fished too heavily. The Peruvian anchovy catch soared, then plummeted, but most scientists still thought the fish would rebound to previous high levels in a short period of time. "In the past decade, the rate at which stocks have become depleted has increased rapidly and the increase in the world catch of fish has slackened off significantly," wrote economist Francis Christie in 1977.[9] But the idea that the ocean itself might get less productive did not seem remotely possible.

"It may be rash to put any limit on the mischief of which man is capable, but it would seem that those hundred and more million cubic miles of water, containing every natural chemical element and probably every group of bacteria, supporting every phylum of animals, moving on the surface from the Equator towards the poles, and returning below, stirred to many fathoms depth by the wind—it would, indeed, seem that here at the beginning and the end is the great matrix which man can hardly sully and cannot appreciably despoil," wrote British fisheries scientist Michael Graham in 1956.[10]

For many U.S. policy makers and scientists, the solutions to third-world hunger lay in the transfer of Western technology and ideology to undeveloped countries, which would then follow the American progression toward capitalism and democracy. As historian Michael E. Latham argues, theorists placed Western industrial capitalist democracies—especially the U.S.—at the apex of the historical scale and stressed that the U.S. could lead less-developed countries through a progression to modernism.[11]

Scientists and policy makers talked about the need for conservation and the word was embedded in postwar policy agreements. But it was a definition rooted in early-twentieth-century progressivism, with its belief that nature was to be conserved for human use.[12] There was no counterargument being made against those who claimed fish only had value if they could be harvested; the evidence for an alternative story did not come about until the 1980s. If fish were not caught, they died, and it was immoral to let fish die when they could contribute to easing hunger problems in a world that was increasingly preoccupied by the question of how to provide enough food for a rapidly growing population.[13] In other words, it was immoral for the people in third-world countries to waste food that could be caught by the industrialized nations—even if most of catch was used to make fish meal and oil, to feed livestock in Europe and North America.

Increasing the catch was extremely successful. It was relatively easy to figure out the best configuration for a net, the temperature at which to store frozen fish, and the best ways to package it for transport and storage. One of the important postwar developments was the transformation of research and development from a haphazard process to a "formidable and growing capacity—a system—for targeting human ingenuity toward the rapid expansion of knowledge and the production of new technologies designed to serve perceived or speculative needs."[14] Federal fishery scientists saw their role as increasing markets for American fishermen, but with the new government research vessels, they were increasingly doing more applied research—figuring out how to fish more effectively, the best temperatures to store fish, and developing more efficient processing equipment. Fishing had mostly been something passed on from older to younger fishermen; now the government was doing the research and passing the information on, hoping to make the fleet more efficient at catching and processing the "underutilized species."

There are few areas in the world's oceans where fishermen have not been able to follow fish. Seabed mapping, global positioning systems, fish-finding electronics, and lighter, stronger nets, have all allowed fishing to penetrate the deepest marine canyons. Massive trawl gear once thought to plow the ocean floor in the same way that a tractor tills a field, has been revealed to have done extensive damage to complex, slow-growing seafloor communities.[15]

Once a technology was introduced, it spread rapidly. One of the most difficult things about fishing with a net was the time it took to bring it back onboard the boat. San Diego tuna man Mario Puretic created a hydraulic power block in 1955, making it easier to pull in seine nets. New synthetic nets were being introduced, much lighter than the old cotton nets and requiring much smaller crews. The nets could be scaled up from the basic net design for sardines to catch large volumes of skipjack and yellowfin.[16] A decade later, superfast refrigeration allowed tuna to be caught anywhere in the world and transported to a cannery where the labor was cheapest.

The Puretic Power Block was soon being produced by MARCO Seattle, a boat-building company. It was marketed worldwide and became the "linchpin in the mechanization of purse seining," according to an undated story from *Fishing News International*.[17] It was introduced into the Puget Sound salmon fishery, where it "increased their catches, halved their crews, and reduced their net hauling time." The block was adopted by the menhaden fishery on the East Coast and in the Gulf of Mexico, then by the California tuna bait boats. MARCO developed deck machinery using winches and hydraulics, adapting them as vessels got larger.

Unique versions of the power block were created to develop specific fisheries, such as the pilchard fishery in South Africa. Ships were designed and built for fishing anchoveta in Peru and Chile. The article estimates that MARCO and its shipyard arm built more than 1,000 purse-seine vessels, ranging from 32 to 221 feet. In the Pacific, the U.S. bait-boat fishery off Central and South America was almost completely replaced by purse seiners. Most of the tuna was taken by Japanese and American boats. Norway, Iceland, Spain, and Portugal used the MARCO system to develop herring and sardine fisheries. Once a technological problem was solved, it could be endlessly modified for different kinds of fish.

The U.S. was one of the nations that allowed foreign boats right back into American waters after establishing 200-mile EEZs. The Soviets had been catching some 20 billion pounds (approximately 9 million metric tons) of fish in areas that were now under federal control. There was a loophole in the law that allowed foreign boats to continue to operate in American waters, but they would have to buy fish caught by American boats. The Soviets and Poland petitioned to continue fishing. They set up an innovative partnership called Marine Resources Company to broker the fish processed by the Soviet factory boats; the company projected gross sales of $33 million in 1984. The new law and the new opportunities it created resulted in an explosion of boat construction, especially on the West Coast, as the U.S. fishing industry prepared to claim the fish resources that the Japanese and Soviet boats had pioneered in the North Pacific Ocean. The foreign ships were phased out by 1993. There were American factory processing ships to take their place.

According to Michael Weber's account, between the late 1970s and early 1980s there was a sharp escalation in the number of new American boats. There were 857 new vessels in the offshore shrimp trawl fleet in the Gulf of Mexico, half of them built with federal loan guarantees. When gulf shrimp catches crashed, the bankrupt boats drifted into other fisheries. Many went to Hawaii and to the West Coast, in Oregon, Washington, and California.[18] Some of them even wound up fishing in Alaska, in the Bering Sea. It became painfully obvious that a boat that sold for pennies on a dollar could be taken to another fishery, obtain a new license (because there were no restrictions on entering new fisheries), and be back fishing again on the same limited stock of domestic fish. The new government and private money Americanized the fisheries, but it could do nothing to reset the clock for the fish. There was a strong strain of individualism within the industry that would fight attempts to limit the number of boats and licenses.

The U.S. signed governing international fisheries agreements (GIFAs) with Poland, Romania, Spain, Bulgaria, the Republic of Korea, the German Democratic Republic, the Republic of China, the Soviet Union, and the European Economic Community between 1977 and 1993 when the foreign boats were finally phased out.[19] Permits were granted to 699 catching vessels and 21 processing vessels; the vessels paid $10 million for the access, with $3 million being refunded if nations could not catch their quota. The agreement with the Soviet Union included language about collaboration on scientific research and the collection of biological information.[20]

The Fisheries Act accomplished in a decade what several decades of financial and technical assistance could not, as Weber noted. "In a matter of years, it enabled many U.S. fishing fleets to become as capable of overfishing as the foreign fleets that were gone from U.S. waters."[21]

Money also flowed in from other countries, especially Norway and Japan. As the 1960s went on, Norwegians became major participants in the Alas-

Fig. 8.1. Nick Bez at dinner with Harry Truman. Permission from Bez family. Nick Bez, the fisherman turned aviation pioneer, was also known for his friendship with President Harry Truman.

kan king crab fishery, investing heavily in new factory trawlers. Several vessels were basically built in Norway and sent to Alaska to fish for king crab and for pollock.[22] Norway's Christiana Bank had a loan portfolio of $435 million in the Seattle-based fishery, according to Kevin Bailey's account of the growth of the pollock fishery. Bailey quotes an American fisherman as saying "you could get into a factory trawler with nothing down."[23] There was also money from the Japanese, most of it invested in fish processing companies in Alaska.[24]

American fish processors who knew what they were doing could make large profits, and Nick Bez certainly did during the 1960s. He sold his stock in Astoria's Columbia River Packers Association in 1951 and bought P. E. Harris, one of the largest salmon-packing and distributing companies in the Northwest, founded in 1916. Bez took over the assets and became president, board chairman, and the largest stockholder. The company had packed salmon under a variety of names, including Peter Pan, Gill Netters Best, and Sea Kist. Bez renamed the company Peter Pan Seafoods, Inc., and it operated canneries in Alaska, Puget Sound, and Astoria at the mouth of the Columbia River. He broadened the marketing line to include canned tuna and Alaska king crab. He was bringing frozen Japanese tuna to the U.S. to can, evading any tariff, and worsening the economic situation of the Southern California boats, an act that undoubtedly gave him great pleasure, seeing as how the

Southern California boats had forced him out of Central America. His was one of the first American companies to import salmon and tuna from Japan to process in the U.S., and he opposed any increase in fees on Japanese tuna.[25]

The *Pacific Explorer* was eventually sold for scrap, but Bez stayed in the news.[26] Two of his boats, the *Western Clipper* and the freezership *Toni B*, were charged with "illegal fishing" in Peruvian waters in 1955. He refused to pay the fines. "We aren't going to recognize any 200-mile limit," he told the *Seattle Times*. "If they get away with 200 miles, they could make it 10,000 miles."[27] The *Toni B*, described as a former navy tug, sank in February of 1955, in "heavy Caribbean waters." Bez's son John was the skipper of the boat. All 10 men were saved when they were picked up by a navy vessel. The *Toni B* had been on her way to deliver 600 tons of tuna to the new cannery at Ponce, Puerto Rico. The cargo was valued at $180,000, according to clippings from several unknown newspapers in the possession of the Bez family.

Bez had always been colorful, but when he was first quoted in the press, his accent was reproduced, with faulty grammar and incorrect syntax. He was painted as an immigrant who made good, a penniless boy who became a millionaire, a smart man with no formal schooling; the coverage was often patronizing. The tone in his hometown paper was a lot more respectful by 1960, when Bez threw a party for 500 at the Rainier Club Lounge. He was celebrating the fiftieth anniversary since his arrival in the U.S. His friends included most of the city's political and financial elite.[28] He was still the friend of presidents, chairing a Seattle committee to raise money for the John F. Kennedy Presidential Library. Nick Bez died in 1969 at the age of 73.

In late 1965, he announced that Peter Pan and Taiyo Gyogyo K. K. had jointly purchased canneries in Dillingham, Naknek, Port Moller, King Cove, and Squaw Harbor. The company would pack salmon and crab, and Bez said that Taiyo would remain in the background, as a stockholder (with 49% of the shares).[29] Taiyo also jointly owned Western Canada Whaling Company with British Columbia Packers, Ltd., and was operating a whaling fleet on the West Coast of Vancouver Island.

A few months earlier, the New England Fish Company announced its Orca cannery would be run jointly with Nichiro Gyogyo Kaisha, Ltd., and Mitsubishi Shoji Kaisha, Ltd. The most significant investment came in 1973, when Kyokuyo Hogei acquired 98 percent of the stock in Whitney-Fidalgo Seafoods, Inc. The Seattle firm was the largest fish processing company in Alaska.[30] The Japanese so dominated Alaskan fish processing that it drew a congressional investigation. The Japanese would finally secure their access to something they had coveted since the 1930s: the rich salmon fisheries of Bristol Bay.

THE SOVIET UNION AND JAPAN

The two largest fishing nations in the world were obviously the worst hit by the changes in the law. Japan and the Soviet Union would each lose about 4 million tons of annual catch, according to a 1979 analysis.[31]

The Soviets owned more than half of the world's tonnage of big fishing vessels, more than 5,000 fishing vessels, factory vessels, and fish transport. It was four times as large as the Japanese, the next largest in size. It was also increasing about 10 percent a year. Bulgaria, East Germany, Poland, and Rumania also sharply increased in their vessel tonnage, all of it for distant-water fishing. The Soviet and East European fishing fleets made up almost 60 percent of the world's total tonnage in fishing vessels more than 100 GRT in size.[32] The Soviet bloc countries were taking between 50 and 70 percent of their catches thousands of miles from home, greatly increasing their harvesting costs. Fuel costs increased steadily through the 1970s, rising to more than 20 percent of harvesting costs.[33]

By the 1960s, the Soviets had trawlers in the waters off Greenland, Labrador, Nova Scotia, the New England and mid-Atlantic states, and south off Campeche, Mexico. Their boats were off Argentina, Uruguay, and southern Brazil, as well as the northwest and southwest coasts of Africa and southern Arabia. Expansion was initially slower in the Pacific, but once fishing began, it was carried out at a faster rate.[34]

The arrival of the Soviet fleet off New England coincided with unusually good conditions for the hatching and survival of haddock (*Melanogrammus aeglefinus*) in the North Atlantic. By 1965, there were small haddock across the North Atlantic and small American trawlers watched helplessly as they were scooped up by the Soviet fleet. Soviet haddock catches increased tenfold in the space of a year—an incredible 249,000 metric tons was caught in one year. East Coast fishermen began to push Congress to do something about the foreign boats. When the haddock catches slowed, the Soviet boats moved on, and the American fleet, which had been catching 62,000 metric tons of haddock in the 1950s and 1960s, caught a mere 3,731 metric tons by 1974.[35]

Author William Warner wrote a powerful book, *Distant Water*, about the factory fleets in the North Atlantic in the 1970s. More than a thousand Western European and Communist-bloc fishing vessels "swarmed across the Atlantic to fish North American waters," Warner wrote. The fleets caught 2.1 million tons of fish, 10 times the New England and triple the Canadian Atlantic catch, in 1974. "Huge as the total catch might seem, the catch per vessel was down and the fish were running generally smaller than before, even though the foreign boats were fishing longer hours with improved methods over a longer range for a greater part of the year."[36] Warner recounts an interview with a Soviet captain, "And then there are all the fish of the deep oceans! Who knows how many new fish and in deeper waters?"[37] By then, the deeper waters in the Pacific, home to rosefish, had already been exhausted.

The presence of the foreign boats has not drawn much scholarly attention, and the single most informative writing that has been done since Warner's book on the expansion of foreign fishing is a novel by Martin Cruz Smith, featuring his Soviet detective, Arcady Renko. Smith introduced the brooding detective in *Gorky Park*, which ended with Renko being declared politically unreliable and fleeing Moscow, chased by the KGB. He wound up in Siberia on a factory ship called the *Polar Star*, working on the slime line, cleaning soft-bodied pollock for freezing and shipping back to the Soviet Union. When a female crewmember is murdered, Renko is forced to investigate his ship and the four American boats (one of which is a Gulf shrimp boat, a very nice touch) that are delivering Alaska pollock to the *Polar Star* for processing. Although published in 1989, the novel is set after 1976 during the joint venture fisheries, where American boats caught the fish and transferred them to the foreign ships for processing. It is not a spoiler to reveal that the Russians were spying, as were the Americans.

The novel captures the danger and the hard labor of the crew, working in cold and wet conditions, on a constantly moving ship in the grey Bering Sea. It is not much different from the conditions of the Japanese crew onboard the *Hakuai Maru* (Brotherly Love), which inspired Takiji Kobayashi to write *The Factory Ship* in 1928. As Renko would have wryly predicted, when the Japanese crew mutinies against the harsh conditions and an Imperial Navy vessel appears, the navy arrests the strike leaders, not the captain.

Japan had the world's second-largest fishing fleet by 1974, about 12 percent of the world's total. It also operated the world's largest fishery, processing 652,000 metric tons of pollock. Nearly 45 percent of Japan's total catch came from the 200-mile EEZs of the U.S., Canada, and the U.S.S.R.[38] The Japanese continued to enter into bilateral negotiations for access to stocks around the world. A 1974 report cited 135 overseas investments with a value of $59 million. Most of the agreements were in Asia and Oceana, but there were also agreements in North America, Africa, and Latin America.[39] Japan was also being increasingly challenged by competition from other longline fleets, especially Taiwan, in both the fishing grounds and the marketplace.[40]

Spanish fishermen who took 70 percent of their catch off the Americas and along the West African coast found themselves having to negotiate for access to fish.[41] As Iceland expanded its seas, fishermen and boats were hurt in the British distant-water fishing ports of Hull, Grimsby, and Fleetwood, where boats lost their access to fish.[42] One of the fastest-growing fisheries had been off West Africa, where European and Asian nations increased the catch from 1.4 million metric tons in 1967 to 3.7 million by 1976.[43] Especially hard-hit were bluefin tuna. Scientists declared the overfishing of bluefin an "international disgrace" in 2008, when a review committee declared the international community deserved better management of the iconic fish.[44]

As nations widened their territorial seas through the 1970s, the costs of fishing were increasing; even the Japanese had to pay more for labor. Many countries eliminated foreign fishing with no compensation, leaving long-range fleets unable to employ their fishermen.

The sudden tightening of regulations sent some Japanese and Soviet vessels fishing in unregulated waters, according to Bailey. There was a large year class of pollock in an area of the Aleutian basin, generally called the Donut Hole, since it is outside the territorial limits of all countries. China had just bought 15 large new factory processing ships and was moving into fishing for pollock. So were the South Koreans. There was no regulation—the pollock catches increased until the stock collapsed in 1994, after harvesting a final catch of 10,000 metric tons.[45] A convention was finally ratified in 1994 but the pollock population in the Donut Hole had been destroyed.

Astoria was at the center of West Coast groundfish research during the 1960s. Scientists continued to build on the foundation established by George Yost Harry, Dayton Lee Alverson, and Jergen Westrheim. The Canadians were building a new state-of-the-art research ship, the research vessel *G. B. Reed*. They wanted Jergen Westrheim to run it and put Canadian fisheries science "on the map," recalled Donald Gunderson. Before he left Astoria, Westrheim reviewed all the voyages the *John N. Cobb* had made during the last decade and plotted the points where rosefish had been found. The Russians were two years ahead of him, but Westrheim was ready to go as soon as the *Reed* was. Their first voyage together was in January of 1963.

Westrheim found that rosefish dominated the Gulf of Alaska west from Cape Spencer, "a huge resource and all of it composed of one year class that hatched in 1953," Gunderson wrote. "He believed that some oceanographic phenomenon had occurred which had transported the larva and young fish to the Gulf of Alaska, where they remained to be harvested by the Soviet fleets which found this vast resource. However, the 1953 year class was actually 10 years old when the Japanese and Russians hit it."[46] Rosefish are slow to mature; the enormous year class was probably caught before they spawned.

Fig. 8.2. Jergen Westrheim was chief scientist onboard the Canadian R/V *G. B. Reed*.

The Soviet boats followed the fish south to the Oregon-California border in 1966, in a fishery that lasted three years before the rosefish were exhausted and the fleet switched to a new target, hake (*Merluccius productus*). West Coast hake turned out to be far more plentiful than rockfish, but it contains a microscopic parasite that softens the flesh as soon as it is caught, limiting the development of a shoreside market.

Even small American trawl boats were landing big loads of the bright red fish. Astoria trawler George Moskovita caught the biggest catch of his life one day in 1965. His boat, the *New Hope*, was running, although Moskovita did not record where the boat was when he saw "a big black spot" on the fathometer. "I got pretty excited because I knew that meant fish. Boy, did it ever! In our first tow we got 50,000 pounds of Perch." By the end of the day, there were 150,000 pounds fish onboard the 84-foot trawler, filling the hatches and piled on the deck "so the boat was nearly sinking."[47]

How many fish could be caught? The Bureau of Commercial Fisheries hired a young biologist from Montana, Donald Gunderson, to run trawl surveys off Washington and Oregon to get an idea of the abundance of the rockfish. Then the Soviets appeared. "Before Gunderson's team of scientists had a chance to figure things out, it was too late," wrote Kevin Bailey, who compared the Soviet harvest of rosefish to clear-cutting a forest of very old trees. The rapid decline of Pacific ocean perch hurt fishermen like Moskovita. "The fish don't stand a chance to survive with this kind of new gear," he wrote.[48] In just two years, his catch per hour would drop by 45 percent.[49] As Captain Gordon White told me decades later, the Rosies were gone.

"Intensive exploitation by man created a sudden change in their population biology, and one that they were poorly adapted to cope with," Gunderson wrote in the summary of his 1975 thesis. "Pacific ocean perch lack the resilience of highly fecund, oviparous groups like the gadoids and their ability to maintain even current levels of abundance is uncertain."[50] Compounding the problem was the work by Westrheim and others, including Gunderson, showing that rockfish were actually far older than scientists had thought.[51] That was ominous because it meant that the fish could only sustain very low fishing levels. And if they were overfished, recovery would be long in coming, if at all.

With the passage of the Fisheries Conservation and Management Act in 1976, Congress created eight regional councils and charged them with writing fishery management plans for the regions they now managed. For the first time, fisheries would be managed regionally, outside three miles. The foreign boats would be phased out and Americans would finally industrialize their fisheries, with generous assistance from the new federal programs to subsidize the Americanization of the fleet. Since the foreign processing ships were still working American waters, the push was to build trawlers to catch fish for delivery to the motherships for processing, as Martin Cruz Smith describes in *Polar Star*. In addition to the new fishery management plans—and this part is very important—the councils were to rebuild the depleted stocks, which would help to revitalize the industry.[52] There was an essential tension between the goals.

Fishery management is often based on "an accumulation, not a selective integration, of different vaguely defined value systems," the byproduct of complex compromises among users, according to biologist Dennis Scarnecchia. Plans can be "surrealistic aggregations of incongruent management goals,

Fig. 8.3. Donald Gunderson was one of the first scientists to study the impact of Soviet fishing on rosefish.

objectives, and actions suggestive of many value systems but truly indicative of none."[53] The objectives of fisheries management are often contradictory, as Daniel Pauly has noted: sustaining harvest, increasing employment and exports, and improving efficiency.[54] Fisheries management is complex, thwarting attempts at change. It is much easier for politicians to support policies for growth, rather than regulations that will put their voters out of business. The more we know what *not* to do, the more we continue to act in ways that we know will ultimately diminish the catch.

This perverse outcome is generally explained as one of the consequences of the "Tragedy of the Commons." Since Garrett Hardin published his essay in 1968, we have tended to understand overfishing in his terms. His famous essay did not mention fishing, but fishing has come to be the exemplar for how tragedy is set in motion: there is no incentive to conserve fish if they can be caught by other fishermen (or, more importantly, by the fishermen of other countries, as this history has shown). At the time Hardin published his essay, the commons was very much under siege, but it was due to the deliberate actions of governments in building boats. Hardin's explanation placed the focus on the individual decisions of fishermen, not on the state policies that created the conditions that sent boats fishing in the first place. Fishing takes place at the will of the state.

Hardin's analysis masked the role of the state in creating the conditions that led to overfishing and, in effect, rewarding fishermen who overfished, at least in the short term. While the Canadians, with their fisheries under federal management, acted swiftly to limit licenses and restrict expansion, the U.S. relied on trip limits, size limits, and restrictions on how often boats could land fish, increasing inefficiency and costs. The frustration over inefficiency in fishing has long been the pet peeve of a generation of economists, who have advocated neoliberal policies of limited private ownership to resolve the tragedy of the commons problem. This misunderstands the true basis for the problem, which is that governments subsidized fishing more for political than for economic reasons. That the fallout from the policy hit individual fishermen was nothing new; the fallout from supporting the economies of Japan and Iceland fell on millions of other American workers.

Hardin was right about something that is overlooked in his famous essay: that it is not possible to maximize more than one variable. In writing of a finite world that can only support a finite population, Hardin points out that Jeremy Bentham's goal of "the greatest good for the greatest number" cannot be realized, partly because of mathematics, but also because "it is not possible to maximize for more than two (or more) variables at the same time."[55]

Postwar optimism and faith in technology suggested that governments could have it all, to expand fishing as a territorial claim, a promise of economic benefits that would come with increased fishing, yet also conservation in perpetuity. Under the Americanization of fishing, a fleet of new boats would replace the foreign fishermen, but stocks would also be rebuilt—an arrangement that is supposed to benefit the government, the fishing industry, fishing communities, and the fish. It should be clear to all that the arrangement has not worked out for the fish.

Modern fisheries management emerges from what historian of science Jennifer Hubbard argues was a "conservation zeitgeist" that included ideas about efficiency, surplus, and abundance, shaping the goals of many postwar sciences. "The argument here is not that the mathematics of MSY was influenced by forestry science," Hubbard writes. "Instead, the philosophy of management was imported from German scientific forestry into fisheries, especially ideas about efficiency."[56] These same ideas also greatly shaped American forestry science, with its emphasis on removing old trees that did not produce good lumber with younger trees that would grow faster. "Senescence" was to be avoided in a properly managed forest and ocean.

During the 1930s, E. S. Russell, the Director of Fishery Investigations for the British Ministry of Agriculture and Fisheries, identified a "special kind of mortality" that fish faced: hunting by humans.[57] Russell argued that human predation is different from natural predation because it targets older and larger individuals. Two decades later, American biologist Milner Bailey Schaefer in effect took a huge step backward when he argued that fishing was just an increase in the predation rate, justifying the "surplus" that could safely be harvested from fish stocks.[58]

"The utilitarian biological, economic and social assumptions which actually underlay both forestry and fisheries science went largely unquestioned until the late 1960s when a new understanding of conservation began to permeate the natural sciences," Hubbard writes. "In fact, the quest for efficiency and the assumptions that fed MSY were so much a part of the political, economic and social thinking of the interwar and postwar eras that they were invisible to the scientists of that era, because they sounded rational, practical, and obvious."[59]

According to Hubbard, A. G. Huntsman delivered a paper about the concept at the 1947 Symposium on Fish Populations in Toronto. O. E. Sette of the U.S. Fish and Wildlife Service took up the concept and it became central to Schaefer's 1954 mathematical treatment of fished populations, his "surplus production model." In dense populations, fish were old and slow-growing, resulting in a small annual crop. Thinning out the old population through intense fishing replaced the old, slow-growing fish with younger, fast-growing individuals, increasing the weight of the crop, just as thinning trees increased the yield in a forest.[60]

As the oceans Balkanized into Cold War camps, so did marine science. Biological oceanography expanded and focused more on basic research about how the ocean functions. The intellectual exchange between oceanographers and fisheries scientists dwindled, and it remains very much that way today. The chasm between these two disciplines is reflected by their present-day reputations. Biological oceanographers are said to study life at the bottom of the ocean's food chain whereas fisheries scientists study animals at the middle and top.[61] Oceanography is also fragmented from within, with the development of biological, chemical, and physical oceanography, as well as marine biology, meteorology, and geophysics. "Oceanography as a discipline still, in some ways, does not quite exist," writes historian of science Naomi Oreskes.[62]

The focus of fisheries science has shifted over the decades, from identification and classification to mathematical models, a trend solidified with the introduction of computers that made it possible to analyze increasing amounts of data.[63] These fish population demographics studies, used to develop fishing equations for predicting stock abundance, became the mainstay of American fisheries biology.[64] But such computer models and population studies do not exist in isolation. They emerge from the conceptual foundation of fisheries science, often invisible, containing components of thinking that we now know to be misleading, like surplus production theory.[65]

The Rome meeting in 1955 established the precedent that fishing could not be regulated until scientific proof of overfishing was established. The burden of proof is on the science, and it is only too easy to criticize the many assumptions that go into the mathematical models. MSY is theoretically the weight of the oldest cohort in a population when it reaches its maximum weight; it thus requires an infinite number of boats.

MSY depends on a large number of estimates, often made with faulty and incomplete data. It assumes that populations reproduce and grow, and that growth rates can be increased by thinning the population, especially the older, slower-growing fish. This produces the "surplus" that can safely be harvested. Scientists have focused on estimating the total biomass of fish, rather than the different sizes and age structure of the population. Many small, low-tropic level species have collapsed with little notice, reducing the food supply for larger fish, seabirds, and marine mammals.[66]

The confusion over MSY extends to scientists as well. In his history of fisheries science, Tim S. Smith writes that three partial theories made up

MSY.[67] He cites the work of British scientists Raymond Beverton and Sidney Holt, Canadian scientist William E. Ricker, and American Milner B. Schaefer. Ricker (1908–2001) published the spawner-per-recruit theory, which outlined means to estimate the optimum number of spawners (or parents) from each year class of fish. Schaefer (1910–70) argued that fish had surplus that could safely be taken by fishermen. Smith contends that these theories together make up MSY, but this is incorrect. The work of Beverton and Holt is considered the intellectual foundation of fisheries science. It emerged from Lowestoft between 1947 and 1957, when *On the Dynamics of Fish Populations* was finally published. It has nothing to do with MSY, which I argue is a political construct, a tool the State Department adopted in its battle to maintain open seas for American military vessels.

According to Sidney Holt, the idea of "surplus production" was implicit in his work with Beverton, as well as within Ricker. It was only Milner Bailey Schaefer who used the term, with its anthropomorphic and industrial implications, that the fish were available to be harvested. Holt calls the adoption of MSY one of the three tragedies that befell international fisheries in the 1950s. The other two tragedies are failing to limit the number of boats, and using surplus production models to provide pseudoscientific advice (Sidney Holt, personal communication).

Did sacrificing the jobs of American fishermen and processing workers help bring about the foreign policy objectives? For Iceland, economic aid, primarily from the U.S., was more than $70 million in grants and loans between 1948 and 1960, increasing the national capita income about 11 percent between 1952 and 1957. "This was the price the United States and NATO paid for political and economic stability, for their military presence, and for limiting the influence in Iceland of the Soviet Union and its supporter, the Socialist Party," concludes historian Valur Ingimundarson.[68] As President Eisenhower ruefully observed, buying the catch directly would have been cheaper.

The U.S.-Iceland trade agreement certainly helped achieve the Cold War goals of the restoration of the Japanese economy and tying the Icelandic economy to the West, by creating markets for Icelandic fish so it would not be sold to the Soviets. "By reorienting Japanese commerce away from the Western Pacific basin, granting its imports liberal tariff treatment, and fighting Western European opposition to Tokyo's accession to GATT, America stimulated a boom in Japanese trade that endured into the 1990s," writes political scientist Thomas Zeiler.[69] Japan was ranked thirteenth in exports to the U.S. in 1952, but within eight years, it was second only to Canada. "Complaints from American producers about the Asian invasion of their markets, and continued Japanese trade discrimination, compelled Tokyo to adopt a comprehensive trade liberalization plan in 1960," wrote Zeiler. "But trade patterns confirmed that Japan had developed into both America's commercial partner and rival."

For the Soviet Union, fishing was a source of desperately needed protein; a way to modernize and industrialize on a massive scale, building the world's largest fishing fleet; and a challenge to the American ocean supremacy by fishing within sight of their land. The fleets were enormously efficient, until nations expanded their territorial seas.

The picture for the Pacific Islands is far more mixed. Just having fish in your backyard is no guarantee that you can catch them and sell them competitively on the world market. There was a window of opportunity after World War II to use fishing as a way to industrialize and modernize, but that window had closed by the late 1980s when the islanders sought to build their own purse-seine fleet. Japan and the U.S. delayed recognizing the islands' exclusive economic zone, allowing new boats from Taiwan and South Korea to enter the purse-seine fishery for tuna. Residents are disappointed that they have not commanded a greater share of tuna resources.[70] The lowering of wages paid to Samoans in the 1950s opened the door to sweatshop labor in textile plants decades later. Today the islands are threatened with rising waters and a need to relocate, a wrenching and expensive proposition. The profits the Japanese, American, South Korean, and Taiwanese tuna fleets have made on Micronesian tuna has been at the expense of the Micronesian people, who continue to watch the fleets of other nations catch the fish they want to be catching themselves.

For the American policy makers, fishing was never more than a political card to be played to achieve foreign policy objectives, and a minor card at that. The consequences of having science shaped by political processes have been terrible for fisheries science and for fish stocks. It was easy to build boats and catch fish, but understanding what was going on in the ocean and within its fish communities has taken a far, far longer time to decipher.

CONCLUSIONS

Updating the Best Available Science

> At the end of a fishing season it is deemed to be a good custom to look over the distance covered and summarize the principal conclusion of one's work. This will be done here.
> —Jon Th Thor[1]

Imperialism is generally defined as a nation taking control of the resources of others and using them to its own benefit. Throughout history fishing has been a strategy of imperialism. Control of ocean space has always been an important component of state power.[2] "The discovery of the sea inaugurated a new age in which control of the world's trade, and to a considerable extent also political control, fell gradually into the hands of a small group of states, mostly in Western Europe, which could build enough reliable ships to operate in all the oceans at once, and move at will from ocean to ocean," wrote maritime historian J. C. Perry.[3]

After World War II, more nations, both old and new, set their sights on the oceans as a way to develop their economies, secure a supply of cheap food, and make territorial claims for reasons of state. Overlying this construction of ocean power was the Cold War, with its ideological battle between communism and capitalism. There has been a great deal of scholarly attention on the impact of the Cold War on oceanography, but less on the construction of another ocean science, fisheries science, and the construction of the concept of best available science in 1976.

This transnational history shows how rapidly postwar capitalism, facilitated by trade agreements and technology transfers, created a global fishing economy along the lines of Immanuel Wallerstein, who published *The Modern World System* in 1974. Wallerstein saw one world, connected by a complex range of economic relationships. There were three interaction spheres:

the core, with access to capital; the peripheries, where natural resources could be found; and the semiperipheries (or semicores), where transactions took place. Fundamental differences in social development facilitated the concentration of power in the global cores, and to a lesser extent, the semiperipheries. The cores are dependent on cheap labor and raw materials from the periphery areas; in return, cores supply finished manufactured goods.[4] The periphery areas are transformed by their incorporation into the core and the world economy.[5] Fish are the ultimate semiperiphery resource, moving seamlessly throughout the oceans, and they were integrated into the core of the world economy through the expansion of capitalism, imposing social change in many countries in the process.

Governments during the Cold War used trade, tariffs, and treaties to reshape international economic relations. The industrialization of fishing resulted in the creation of a great deal of wealth, as well as great poverty among groups of workers, on boats and in processing plants. Fishermen and fishing communities have traditionally been poor. Sailors and fishermen in the Dutch Golden Age lived on "the edge of bare subsistence."[6] The Meiji restoration created modern Japan and a key factor was the exploitation of the working masses—in mines, factories, and, as we have seen, on fishing boats, with "the legal privileging of the exploiters over the exploited in every way," as Jon Halliday has written.[7] More recently, the *New York Times* reported on fishing operations in Thailand that use slave labor to catch tuna that is fed to pets and livestock.[8]

At the end of World War II, fishing became an important industrialization strategy that required government support, not only in an initial stream to build a fishing fleet, but ongoing support to achieve associated policy goals: cheap food, local employment, territorial claims, or to challenge a world power. With each improvement in fishing equipment—from sail to steam, steam to gas motors, then to diesels, and on to industrial fishing fleets—the intensity of fishing increased, leading to declining catches on the home grounds and forcing boats to fish further from home, increasing their costs.[9] Fishing has always had to be sustained by finding new stocks to fish to make up for declines in the home grounds. This strategy can only be sustained as long as there are new stocks remaining to be "found."

This historical analysis shows that fishery science was shaped by political and economic forces at the national and international level. The oceans were one of the most important and dangerous battlefields of the Cold War. Seeing fish as a renewable resource helped justify the industrialization of fishing. Fishing staked a claim, and as the Cold War deepened, fish, fishermen, and fishing fleets quickly became chess pieces in more complicated games. The resulting policies had little to do with "conserving" fish, as we understand the idea today.

It is only in retrospect, through the study of history, that we can see the ideological construction of postwar science and its emphasis on surplus stocks and harvesting older fish. These ideas about efficiency and use reflected the growing industrialization of nature after 1920, as scientific concepts of industrial management were applied to natural resource systems in an effort to standardize and increase production.[10] Arthur McEvoy argues that by the 1930s, scientists were familiar with the sustained yield concept, even if the details were not fully developed. The concepts were also reflected in the statutory mandates of the U.S. Forest Service and the Bureau of Land Management.[11] "Surplus" fish could be harvested. Salmon hatcheries promised to produce as many fish as needed and the demand, of course, was unlimited.[12] The language adopted at the Rome meeting in 1955 endorsed human actions to increase fishing stocks.

The "science" pushed by the State Department was that as older, slower-growing fish were taken, there was more feed for younger fish to grow faster, bringing the ocean back into equilibrium. Stocks had surplus, and if fishing became unprofitable, fishermen could pause while the stocks rebuilt. It is certainly implied that stocks would rebuild swiftly, and perhaps they would have, if ocean productivity had not declined as much as it has during the last century.

As fishery scientists focused on estimating the biomass in the ocean—and how much of it could be safely harvested—other scientists started to look at the population structure of the fish, and began to realize that it was the larger and older fish, especially the larger and older females, that played the most important role in recovery from disturbance. They produce larger and more viable progeny, a finding that was realized when old females disappeared from the West Coast rockfish landings, setting off the West Coast ground fish fishery disaster in 2000.[13]

Throughout 1960s and 1970s, scientists briskly turned out papers estimating what the global fish catch could be, and what could be sustained. With the fish catch increasing at some 8 percent a year throughout the 1960s, they tried to estimate how high the global catch could go. John Gulland's 1970 estimate of 100 million metric tons was roundly criticized as being too conservative.[14] For most of the last decade, the peak ocean catch was estimated to be around 86 million metric tons, caught in 1996. However, a new reconstruction by Daniel Pauly and Kirk Zeller suggests the peak catch was actually 130 million metric tons, based on an analysis that includes data from sports fishing, bycatch, and the catches of small artisanal fisheries. If the reconstruction is correct, then the decline in global catch is even more significant and steeper than scientists have thought.[15]

Despite all the attention to the problems of the ocean and its fish, biologist Daniel Pauly does not think there is a comprehensive understanding of just how serious the crisis really is. The world catch is declining while effort is increasing; as fishing becomes less profitable, boats seek deeper water and more distant grounds. New species of fish are renamed to make them more appealing to consumers, who eat the fish thinking that fishing is sustainable. The system survives through government subsidies that keep building boats and subsidizing their operations.

"The public at large . . . does not seem to be aware of the damage that subsidized industrial fishing inflicts on marine ecosystems," Pauly wrote, adding that many fisheries scientists "are generally not aware that fisheries are now part of a global system, and that the forces shaping them are not necessarily visible through the study, however detailed, of one fishery on one fishing ground in one country."[16]

When we think of fishing, it is the iconic image of the lone fisherman in his boat that comes to mind—man against the sea, living on his wits like George Moskovita. Dramatic and evocative images of historic fishing make it easy to romanticize what was always a hard, brutal, and dangerous job, an extractive industry that takes place out of sight of land. The real power within the fishing industry is invisible and intangible, the relationships between the processors and the manufacturing companies, allied with government, creating policies focused on maximizing the harvest. In the process, ecosystem function has been compromised, weakening the stocks as they respond to always changing ocean conditions, now exacerbated by climate change.

Our understanding on these issues tends to be local, not global. Three Northwest biologists, Willa Nehlsen, Jack Williams, and Jim Lichatowich, shifted the debate on wild salmon in the Pacific Northwest when they published their groundbreaking paper in 1991. It identified 212 salmon stocks at risk across four states. Almost overnight, as the paper was publicized in the media, that debate shifted from the status of a single stock in a stream to a much wider framework and a realization that salmon decline was more serious that most of us believed.[17] Ransom Myers and Boris Worm wrote in 2003 that industrial fishing had removed as much as 90 percent of the largest fish in the sea.[18] Published in the magazine *Nature*, the paper received a lot of public attention—much of it critical, especially from fishery scientists, who complained both about the conclusion as well as the methodology. But with the publication of the paper, the decline of large, old fish in the ocean, and the global fisheries crisis more generally, penetrated the public consciousness and the disputes among the scientists became front-page news. Our understanding of global fishing shifted from the impact of fishing on single stocks to a more comprehensive view, with scientists soon after calling for a shift in perspective and practice from a single species to an ecosystem approach.[19] The local is dependent on the global, as the story of the destruction of rosefish shows.

The relationship between the nation-state and its fishermen shifts over time. The industrialization of fishing, starting with steam engines in fishing boats in the 1880s, made fishing a more important component of domestic economies. Steam trawlers caught more fish, but the local stocks were depleted and boats were forced to find new water to fish. Fishing had to take place on an industrial scale to cover the increased costs. For Northern European fleets, this meant fishing off Norway, Iceland, and Greenland. The development of marine refrigeration in the 1930s meant boats could stay longer at sea and fish further from home. Fishing became more useful to some governments. Japan and Iceland subsidized the development of marine refrigeration for their fleets in the 1930s; both governments saw fish as playing an important role in the national economy.

World War II elevated the importance of the fishing industry in most national economics. Fishing was a vital industry, supplying much-needed protein. Military contracts gave fishermen stable markets, some for the first time. Government money expanded fisheries, creating employment in coastal areas and providing fish for export. Fishery scientists sought to make fishing gear more efficient, food technologists perfected preservation techniques, and home economists showed how to easy fish was to cook.

Our human understanding of the ocean has changed radically since 1945, when the rapid expansion of fishing sparked the territorial conflicts that led to the Law of the Sea process, the creation of the EEZs, and modern fishery and whaling management institutions.[20] It was easy to catch fish and whales, at least initially, and scientists were confident that they stood on the verge of not only understanding the oceans, but also controlling them, using fish to help solve the political problems caused by global hunger. As Dayton Lee Alverson wrote near the end of his 50-year career, he moved "from [his] early belief that the ocean could easily feed the world's hungry (1950s) to [his] current concern about the consequences of overfishing and the ecological changes caused by fisheries all over the world (1990s–forward)."[21]

Over the last few decades, science has shifted from believing there was a balance within nature to seeing a dynamic ecosystem that is marked by complexity and uncertainty.[22] "Ecosystems are so complicated that we are unable to precisely control or manage them," writes fisheries scientist Dan Bottom. "Their components are tightly coupled, their behavior depends on an entire history of unknown events, and they do not respond to disturbance in a simple linear way."[23]

Removing all the old fish will allow the population to reach its greatest natural rate of increase, wrote Spencer Apollonio and Jacob Dykstra in their 2010 exploration of the New England Fishery Management Council, which they called the poster child for fisheries mismanagement. Both authors have long experience with the council: Dykstra was involved in its creation and was a member for seven years, while Apollonio is a marine biologist who has worked for a state management organization and the first director of the New England council. The greatest natural rate of increase is not sustainable. The younger fish population may begin to oscillate, shifting the population into "a lower hierarchical level with inherently faster dynamics and greater instability and unpredictability."[24] Multiple year classes lend stability to fish populations, allowing them to withstand disturbances, such as shifts in currents, temperatures, and food supply. Species that are long lived, with delayed maturity, naturally select for a population structure with multiple age classes. MSY, by reducing the number of year classes, works against evolutionary adaptation to the environment.

It has been 40 years since the passage of the 200-mile bill. The goals and the science have both been tightened in the last three decades; however, the focus is still on harvest, not on sustaining the population structure of the stocks. Stock assessment and management are still overwhelmingly single-species focused, though progress is gradually being made. It is still up to the scientists to prove that overfishing exists before fishing can be restricted. For most fisheries, there are still too many boats fishing on the same limited pool of domestic fish. The institutions and polices that so rapidly built the postwar industrial fisheries are still mostly in place. Governments still have low-interest loans to build fishing boats. The U.S. is currently assisting in the construction of a new fleet of boats to fish for halibut in Alaska.[25] Subsidies, once created, are difficult to unwind. The subsidies may have created the problems that fishery management is now trying to solve, according to work by three Swedish researchers. Sweden's national postwar goals of making the industry competitive, providing cheap food to the population, and ensuring a good living for fishermen were good goals, but it has taken a very long time to recognize how incompatible those goals are with the dynamic and unpredictable nature of marine ecosystems.[26]

There are countries where fisheries are being managed (and most American and European fisheries fit into that category), but it is still important to look at what is happening globally, and to examine the appropriateness of the conceptual foundation of harvest that underlies policy. This historical analysis makes clear that fisheries science was shaped by political goals.

In his study of the geopolitics around whaling, historian Kurkpatrick Dorsey concludes that the story of whaling leaves the reader "grasping for something good in an otherwise gloomy story of decline."[27] The picture with fishing is much brighter, because it is possible to rebuild at least some fish stocks. In the decades since marine reserves were created in the oceans, even though the reserves are small and fragmented, scientists have been studying how the reserves affect fish populations and the communities on which they depend.

It is easy to catch fish. It has been much, much harder to figure out what fishing means for fish populations. It took the development of new technologies in the 1980s to produce the tools that uncovered the growth rings that led to ageing of rockfish. It took until 1996 to realize the outsized role of old fish in the population structure. Another decade passed before the recognition that fish are not islands unto themselves but interact with one another and with their habitat and environment. And it has taken decades for the tactic designed to reverse overfishing—marine protected areas (MPAs), where fishing is not allowed—to demonstrate that it is possible to rebuild ecosystems and fish populations.

Starting in the 1970s, nations have created small zones in the ocean where no fishing is allowed. Virtually every country with a marine coastline has declared one or more MPAs. There are more than 11,000 MPAs in existence; collectively they comprise only 3.5 percent of the ocean surface.[28] It is only recently that the science around MPAs can be considered mature. But there is also substantial social science data showing that it is possible for people to come together and create governing structure to make reserves work for fishermen, fishing communities, and for the fish. MPAs are controversial because they mean less fish for fishermen to divide. They also require highly local political and economic structures that will allow MPAs to do the job they are intended to do: rebuild the ecosystem and bring back the old fish that are so important to sustainability.

At the United Nations Sustainable Development Summit on September 25, 2015, world leaders adopted the 2030 Agenda for Sustainable Development, which includes a set of 17 goals to end poverty, fight inequality and injustice, and tackle climate change by 2030. Sustainable Development Goal 14, Target 5, states "By 2020, conserve at least 10 percent of coastal and marine areas, consistent with national and international law and based on the best available scientific information."[29]

The goal has been adopted as scientists and social scientists are increasingly clear that stocks can be rebuilt, and that communities can shift to sustainable models that will include both fishing and the current understanding of conservation. The evidence comes from the study of MPAs that have been established since the 1970s. Scientists are finding that fully protected MPAs with strong enforcement almost always result in more fish, more species, and, most importantly, larger sizes. Networks of reserves that extend from shallow into deeper waters can protect more biodiversity, since many species move among habitats during their life cycles. These connected networks can protect species while allowing some extractive use between reserves.

The data is now so compelling that 25 scientists met in Rome during March of 2016 to chart a way forward to reaching the UN goals. The 10x20 Initiative held a dialogue between scientists and policy makers about MPAs. Establishing a network of MPAs throughout at least 10 percent of coastal and marine areas by 2020 would help to conserve and restore marine biodiversity and assist in regenerating wild fisheries, according to Ellen Pikitch, the fishery scientist who chaired the meeting. It is possible to bring back the old fish.[30]

Rebuilding global fish stocks will have to happen in a holistic way, with changes in the way people fish, the way that fishing is managed, and in the role of fishing within communities. Just as governments funded the expansion of fishing after World War II—for a variety of socially beneficial goals—so now must governments fund the transition to a system that will support sustainable fishing and communities. Change will not come from within the fishing industry or the fisheries management sector. It will have to come from people outside of the industry, just as activists supported the establishment of the first MPAs. But the first step in the creation of a new way of fishing and management is to start with a new story, about what might be possible.

"Instead of positivistic science that assumes the world is predictable and controllable, we are emphasizing living with uncertainty," writes Fikret Berkes. "Instead of reductionism that seeks to model, for example, individual fish species and fishing fleets separately, we emphasize holistic approaches that consider fisher-fish-environment together."[31]

Our knowledge about the ocean and fish populations has changed greatly since 1945, and even since 1976, when the "best available information" standard was created in the law to make sure that fisheries would be managed by science. Nothing was said about updating the scientific information as new knowledge was uncovered.

The first groundfish management plan was adopted in 1981. POP, as Gordon White's rosefish was now rechristened in the scientific literature, was placed on a rebuilding schedule, which meant fishermen were not allowed to target them. But trawlers still caught some when fishing for other species. By 2000, scientists realized the population was not rebuilding and once again tightened the regulations. The recovery date for rosefish was originally 2016. It has since been pushed out to 2051.[32]

NOTES

INTRODUCTION

1. Hilary Stewart, *Looking at Indian Art of the Northwest Coast* (Vancouver, BC: Douglas and McIntyre, 1979).

2. D. Schimel, C. Redman, J. Dearing, L. Graumlich, R. Leemans, C. Crumley, K. Hibbard, W. Steffen, R. Costanza, "Evolution of the Human-Environment Relationship," *The Encyclopedia of Earth*, May 2, 2007, updated May 3, 2013, accessed February 2, 2015, http://www.eoearth.org/view/article/152703/.

3. John R. McNeill, *Something New under the Sun: An Environmental History of the Twentieth-Century World* (New York: W. W. Norton, 2000), 3.

4. Naomi Oreskes, "Scaling Up Our Vision," *ISIS* 105, no. 2 (June 2014): 384.

5. Daniel Pauly and Dirk Zeller, "Catch Reconstructions Reveal that Global Marine Fisheries Catches Are Higher than Reported and Declining," *Nature Communications* 7 (January 19, 2016), http://www.nature.com/ncomms/2016/160119/ncomms10244/full/ncomms10244.html, doi:10.1038/ncomms10244.

6. The World Bank, *Sunken Billions: The Economic Justification for Fisheries Reform*, accessed June 8, 2014, http://siteresources.worldbank.org/EXTARD/Resources/336681-1224775570533/SunkenBillionsFinal.pdf.

7. Anya von Moltke, *Fisheries Subsidies, Sustainable Development, and the WTO* (London: Earthscan Books, 2011), 2.

8. Elizabeth Mancke, "Early Modern Expansion and the Politicization of Oceanic Space," *American Geographical Society* 89, no. 2 (April 1999): 234.

9. Melvyn P. Leffler, "National Security and U.S. Foreign Policy," in *Origins of the Cold War: An International History*, ed. Melvyn P. Leffler and Davis S. Painter (New York: Routledge, 1994), 37.

10. Thomas G. Paterson, *On Every Front: The Making and Unmaking of the Cold War* (New York: W. W. Norton, 1992), 71.

11. Sayuri Shimizu, *Creating People of Plenty: The United States and Japan's Economic Alternatives, 1950–1960* (Kent, OH: Kent State University Press, 2001).

12. Naomi and Ronald Rainger Oreskes, "Science and Security before the Atomic Bomb: The Loyalty Case of Harold U. Sverdrup," *Studies in the History and Philosophy of*

Modern Physics 31, no. 3 (2000): 309–69. Jacob Darwin Hamblin, *Oceanographers and the Cold War: Disciples of Marine Science* (Seattle: University of Washington Press, 2005).

13. Leon Herman, "The Postwar Expansion of Russia's Fishing Industry," United States Senate Committee of Commerce, January 23, 1964, Fisheries Research Institute, University of Washington.

14. A. T. Pruter, "Development and Present Status of Bottomfish Resources in the Bering Sea," *Journal of the Fisheries Research Board of Canada* 30, no. 12, part 2 (1973): 2373–85.

15. Sarah Keene Meltzoff and Edward Lipuma, "The Troubled Seas of Spanish Fishermen: Marine Policy and the Economy of Change," *American Ethnologist* 13, no. 4 (November 1986): 681–99.

16. William Warner, *Distant Water: The Fate of the North Atlantic Fisherman* (Boston: Little, Brown, 1977), 287.

17. Hiroshi Kasahara, "The U.S. Fishing Industry and Related International Activities," in *The Future of the Fishing Industry of the United States*, ed. DeWitt Gilbert (Seattle: University of Washington Press, March 25–27, 1968), 241–58.

18. Gregory Cushman, "The Lords of Guano: Science and Management of Peru's Marine Environment, 1800–1973" (PhD thesis, University of Texas at Austin, 2003).

19. Chris Reid, " Potential for Historical-Ecological Studies of Latin American Fisheries," in *The Exploited Seas: New Directions for Marine Environmental History*, ed. Poul Holm, Tim D. Smith, and David J. Starkey (St. John's: International Maritime Economic History Association and the Census for Marine Life, 2001), 141–66.

20. *Canadian Fisherman* (June 1949): 8.

21. Department of Fisheries Canada and the Fisheries Research Board, *The Commercial Fisheries of Canada* (Ottawa, 1956), 120.

22. William B. Schrank, "The Failure of the Canadian Seasonal Fishermen's Unemployment Insurance Reform During the 1960s and the 1970s," *Marine Policy* 22, no. 1 (1998): 67–80.

23. A. M. Springer, J. A. Estes, G. B. van Vliet, T. M. Williams, D. F. Doak, E. M. Danner, K. A. Forney, and B. Pfister, "Sequential megafaunal collapse in the North Pacific Ocean: An ongoing legacy of industrial whaling?" *Proceedings of the National Academy of Sciences of the United States of America* 100, no. 21 (October 14, 2003): 12223.

24. Kevin Bailey, *Billion-Dollar Fish: The Untold Story of Alaska Pollock* (Chicago: University of Chicago Press, 2012), 119.

25. Ransom A. Myers and Boris Worm, "Rapid Worldwide Depletion of Predatory Fish Communities," *Nature* 423 (2003): 280–83; Peter Ward and Ransom A. Myers, "Shifts in Open-Ocean Fish Communities Coinciding with the Commencement of Commercial Fishing," *Ecology* 86, no. 4 (2005): 835–37.

26. http://carmelfinley.wordpress.com/2011/01/27/oregon-trawlers-called-them-rosies/

27. Box 76, Folder "Panama Correspondent Sullivan, Paul A.," American Tuna Association files, SIO.

28. Wilbert Chapman, "United States Policy on High Seas Fisheries," *Department of State Bulletin* 20, no. 498 (January 16, 1949): 67–80.

29. J. L. Kask, "Dedication," in *World Fisheries Policy: Multidisciplinary Views*, ed. Brian J. Rothschild (Seattle: University of Washington Press, 1972), vii.

30. Owen S. Hamel and Kotaro Ono, "Stock Assessment of Pacific Ocean Perch in Waters off of the U.S. West Coast in 2011," Pacific Fishery Management Council, Portland, September 20, 2011, http://www.pcouncil.org/wp-content/uploads/Pacific_Ocean_Perch_2011_Assessment.pdf.

CHAPTER ONE

1. Chapman to George H. Owens, September 24, 1949, Chapman papers, 1852, Box 15, Folder Number 23, UWSC.

2. *Pacific Fisherman*, September 1945.

3. *Seattle Post Intelligencer*, June 23, 1945.

4. "Report of the Alaska Crab Investigations," *Fishery Market News* 4, no. 5 (May 1942): 3–8.

5. Philip E. Chitwood, "Japanese, Soviet, and South Korean Fisheries off Alaska, Development and History through 1966," U.S. Fish and Wildlife Service Circular 310, Washington, D.C. (1969): 1–37.

6. *Pacific Fisherman* (September 1937): 27.

7. Harry N. Scheiber, "Origins of the Abstention Doctrine in Ocean Law: Japanese-U.S. Relations and the Pacific Fisheries, 1937–1958," *Ecology Law Quarterly* 16, no. 1 (January 1989): 32.

8. Ann Hollick, *U.S. Foreign Policy and the Law of the Sea* (Princeton, NJ: Princeton University Press, 1981), 19.

9. Donald Cameron Watt, "First steps in the enclosure of the oceans," *Marine Policy* (July 1972): 221.

10. Harry N. Scheiber, *Inter-Allied Conflicts and Ocean Law, 1945–53: The Occupation Command's Revival of Japanese Whaling and Marine Fisheries* (Taipei, Taiwan: Institute of European and American Studies, Academia Sinica, 2001).

11. Charles L. Mee, *Meeting at Potsdam* (New York: M. Evans, 1975), 314.

12. Sayuri Shimizu, *Creating People of Plenty: The United States and Japan's Economic Alternatives, 1950–1960* (Kent, OH: Kent State University Press, 2001).

13. Charles B. Selak Jr., "Recent Developments in High Seas Fisheries Jurisdiction under the Presidential Proclamation of 1945," *American Journal of International Law* 44, no. 4 (1950): 673.

14. Thomas G. Paterson, *On Every Front: The Making and Unmaking of the Cold War* (New York: W. W. Norton, 1992), 80–81.

15. Hal M. Friedman, "The Beast in Paradise: The United States Navy in Micronesia, 1943–1947," *Pacific Historical Review* 62, no. 2 (May 1993): 173–95.

16. Ibid., 183.

17. Harry N. Scheiber, "Pacific Ocean Resources: Science, and Law of the Sea: Wilbert M. Chapman and the Pacific Fisheries, 1945–70," *Ecology Law Quarterly* 13, no. 3 (September 1986): 390.

18. L. Larry Leonard, *International Regulation of Fisheries* (New York: Columbia University Press, 1944), 124.

19. Ward T. Bower, "Alaska Fisheries and Fur Seal Industries in 1936," *Administrative Report No. 28*, U.S. Bureau of Commerce, Bureau of Fisheries, 1937.

20. Charles Gilbert and Henry O'Malley, "Special Investigations of Salmon Fishery in Central and Western Alaska," *Alaska Fisheries and Fur Industries*, Document 891 (1919): 143–60.

21. "Report of the Alaska Crab Investigation," *Fishery Market News* 4, no. 5 (May 1942): 3–10.

22. "The Alaskan King Crab," *Fishery Market News* (May 1942, supplement): 27.

23. Raymond McFarland, *A History of the New England Fisheries* (New York: University of Pennsylvania, 1911), 19.

24. Samuel E. Morison, *The Maritime History of Massachusetts, 1983–1860* (Boston: Houghton Mifflin, 1941), 137.

25. Eugene Skolnikoff, *The Elusive Transformation: Science, Technology, and the Evolution of International Politics* (Princeton, NJ: Princeton University Press, 1993), 47.

26. M. Judd Harmon, "Some Contributions of Harold L. Ickes," *Western Political Quarterly* 7, no. 2 (June 1954): 243.

27. Harold L. Ickes, "The Nation Turns to its Fisheries," *Pacific Fisherman* (January 1943): 27.

28. "A Meeting of the Coordinators Fishery Industry Consultants" (Washington, D.C.: Department of the Interior, February 2, 1944), 12.

29. Ibid., 19.

30. Ibid., 108.

31. Ibid., 100.

32. *Seattle Times*, March 23, 1945.

33. *Pacific Fisherman* (July 1945): 39.

34. *Seattle Times*, March 23, 1945.

35. Box 4, Folder 12, Chapman papers, UWSC.

36. *Seattle Post-Intelligencer*, April 25, 1947.

37. For more on Chapman, see Harry N. Scheiber, "Origins of the Abstention Doctrine in Ocean Law: Japanese-U.S. Relations and the Pacific Fisheries, 1937–1958," *Ecology Law Quarterly* 16, no. 1 (January 1989): 23–101.

38. Wilbert McLeod Chapman, "Observations on Tuna-like Fishes in the Tropical Pacific," *California Fish and Game* 32, no. 4 (December 2, 1946): 167.

39. Box 50, Folder 21, Magnuson papers, UWSC.

40. Box 4, Folder 12, Chapman papers, UWSC.

41. "Fishing vessels returned by Navy," *Fishery Market News* (April 1943): 13.

42. "Construction of 361 new fishing vessels authorized," *Fishery Market News* (October 1943): 25.

43. "OCF approved $3,500,000 in fishery construction," *Fishery Market News* (July 1944): 17.

44. Christopher Howe, *The Origins of Japanese Trade Supremacy: Development and Technology in Asia from 1540 to the Pacific War* (Chicago: University of Chicago Press, 1996).

45. Kurkpatrick Dorsey, *The Dawn of Conservation Diplomacy: U.S.-Canadian Wildlife Protection Treaties in the Progressive Era* (Seattle: University of Washington Press, 1998), 144.

46. J. C. Perry, *Facing West: Americans and the Opening of the Pacific* (Westport, CT: Praeger, 1994), 170.

47. E. G. Mears, *Maritime Trade of Western United States* (Stanford: Stanford University Press, 1935), 389.

48. Ibid.

49. Ibid., 385.

50. Perry, *Facing West*, 172.

51. Jon Halliday, *A Political History of Japanese Capitalism* (New York: Pantheon Books, 1975), 58.

52. Walter Le Feber, *The Clash: U.S.-Japanese Relations through History* (New York: W. W. Norton, 1997), 43.

53. C. Groot and L. Margolis, *Pacific Salmon: Life Histories* (Vancouver: UBC Press, 1961).

54. F. A. Nicholson, *Note on Fisheries in Japan* (Madras, OR: Superintendent, Government Press, 1907), 40.

55. "Japan's Fisheries Industry 1939," *Japan Times and Mail* (1939), 21.

56. Mark R. Peattie, "Nanshin: The 'Southward Advance,' 1931–1941, as a Prelude to the Japanese Occupation of Southeast Asia," in *The Japanese Wartime Empire, 1931–1945*, ed. Ramon H. Myers, Peter Duus, and Mark R. Peattie (Princeton, NJ: Princeton University Press, 1996), 201.

57. Georg Borgstrom, *Japan's World Success in Fishing* (London: Fishing News Books, 1964), 253.

58. *Japan Times and Mail* (1939): 82.

59. George L. Small, *The Blue Whale* (New York: Columbia University Press, 1972), 152.

60. *Japan Times and Mail*, 82.

61. Kurkpatrick Dorsey, *Whales and Nations: Environmental Diplomacy on the High Seas* (Seattle: University of Washington Press, 2013), 81.

62. Small, *Blue Whale*, 152.

63. Greg Dvorak, "Connecting the Dots: Teaching Pacific History in Japan from an Archipelagic Perspective," *Journal of Pacific History* 46, no. 2 (September 2011): 243.

64. Ibid., iv.

65. Sandra M. Tomkins, "The Influenza Epidemic of 1918–10 in Western Samoa," *Journal of Pacific History* 27, no. 2 (December 1992): 181–97.

66. James Joseph, Witold Klawe, and Pat Murphy, *Tuna and Billfish: Fish Without a Country* (La Jolla: The Inter-American Tropical Tuna Commission, 1988), 6–10.

67. Gary D. Sharp, "Tuna Oceanography—An Applied Science," in *Tuna: Physiology, Ecology, and Evolution*, ed. Barbara A. Block and E. Donald Stevens (San Diego: Academic Press, 2001), 355.

68. Perry, *Facing West*, 290.

69. Roger W. Gale, *The Americanization of Micronesia: A Study of the Consolidation of U.S. Rule in the Pacific* (Washington, D.C.: University Press of America, 1979), 22.

70. M. R. Peattie, "Introduction," in *The Colonial Japanese Empire, 1895–1945* (Princeton, NJ: Princeton University Press, 1984), 6.

71. Fukuzo Nagasaki, "Some Japanese Far-Sea Fisheries," *Washington Law Review* 43, no. 1 (October 1967): 205.

72. John G. Butcher, *The Closing of the Frontier: A History of the Marine Fisheries of Southeast Asia, 1850–2000* (Singapore: Institute of Southeast Asian Studies, z2004), 156.

73. Takiji Kobayashi, *The Factory Ship and the Absentee Landlord,* trans. Frank Motofuji (Seattle: University of Washington Press, 1973), 58.

74. Walter A. McDougall, *Let the Sea Make a Noise: A History of the North Pacific from Magellan to MacArthur* (New York: Harper Collins, 1992), 507.

75. Peattie, *Colonial Japanese Empire,* 181.

76. Tomkins, "The Influenza Epidemic of 1918–10 in Western Samoa," 197.

77. Hermann Joseph Hiery, *The Neglected War: The German South Pacific and the Influence of World War I* (Honolulu: University of Honolulu Press, 1995), 243.

78. E. C. Weitzell, "Resource Development in the Pacific Mandated Islands," *Journal of Land & Public Utility Economics* 22, no. 3 (August 1946): 201.

79. Bruce Cumings, "Japan and the Asian Periphery," in *Origins of the Cold War: An International History,* ed. Melvyn P. Leffler and David S. Painter (London: Rutledge, 1994), 219.

80. Gale, *Americanization of Micronesia,* 37.

81. Hiery, *Neglected War,* 131.

82. Weitzell, "Resource Development in the Pacific Mandated Islands," 205.

83. Gale, *Americanization of Micronesia,* 37.

84. Kamakichi Kishinouye, "Contributions to the Comparative Study of the So-Called Scombroid Fishes," *Journal of the College of Agriculture, Imperial University of Tokyo* 8, no. 3 (1923): 293–475.

85. Box 38, Folder 1, "The Tuna Fishery in the Mandated Islands," by M. B. Schaefer, December 1946, Chapman papers, UWSC.

86. Ronald A. Helin, "Soviet Fishing in the Barents Sea and North Atlantic," *Geographical Review* 54, no. 3 (July 1964): 386–408.

87. Elizabeth A. Dunning, "Soviet Distant-Water Fishing after Extended National Jurisdiction" (master's thesis, University of Washington, 1984), 19.

88. H. Kasahara and W. Burke, *North Pacific Fisheries Management* (Washington, D.C.: Resources for the Future, 1973).

89. A. T. Pruter, "Soviet Fisheries for Bottomfish and Herring off the Pacific and Bering Sea Coasts of the United States," *Marine Fisheries Review* 38, no. 12 (December 1976): 3.

90. Dunning, "Soviet Distant-Water Fishing after Extended National Jurisdiction," 19.

91. Paul R. Josephson, *Would Trotsky Wear a Bluetooth: Technological Utopianism under Socialism, 1917–1989* (Baltimore: Johns Hopkins University Press, 2010), 197.

92. Loren R. Graham, *The Ghost of the Executed Engineer* (Cambridge, MA: Harvard University Press, 1993), 90.

93. Roger Munting, *The Economic Development of the USSR* (New York: St. Martin's Press, 1982), 64.

94. J. R. McNeill and Corinna R. Unger, eds., *Environmental Histories of the Cold War* (Cambridge: Cambridge University Press, 2010), 6.

95. Josephson, *Would Trotsky Wear a Bluetooth*, 204.

96. George A. Taskin, "The Falling Level of the Caspian Sea in relation to the Soviet Economy," *Geographical Review* 44, no. 4 (October 1954): 508–27.

97. Ibid., 516.

98. George A. Taskin, "The Soviet Northwest: Economic Regionalization," *Geographical Review* 51, no. 2 (April 1961): 214.

99. Ole E. Mathisen and Donald E. Bevan, *Some International Aspects of Soviet Fisheries* (Cleveland: Ohio State University Press, 1968), 19.

100. Yulia V. Ivashchenko, Phillip J. Clapham, and Robert L. Brownell Jr., "Soviet Illegal Whaling: The Devil and the Details," *Marine Fisheries Review* 73, no. 3 (2011): 3.

101. P. A. Moiseev and I. A. Parakostzov, "Some Data about the Ecology of Rockfish of the Northeast Part of the Pacific Ocean," trans. Donald Bevan, *I Voprosy Ikhtiologii (Problems of Ichthyology)* 1, no. 1 (1961).

102. Stephen Brain, "The Great Stalin Plan for the Transformation of Nature," *Environmental History* 15 (October 2010): 680.

103. Philip P. Micklin, "Desiccation of the Aral Sea: A Water Management Disaster in the Soviet Union," *Science, New Series* 241, no. 4870 (September 2, 1988): 1170–76.

104. Ibid., 1173.

105. Geir Lundestad, "Empire by Invitation? The United States and Western Europe, 1945–1952," *Journal of Peace Research* 23, no. 3 (September 1986): 263–77.

106. *Commercial Fisheries Review*, November 1946, 46–47.

107. Ibid.

108. Ivashchenko, "Soviet Illegal Whaling," 5.

109. Pruter, "Soviet Fisheries for Bottomfish and Herring off the Pacific and Bering Sea Coasts of the United States," 6.

110. Terence Armstrong, "Soviet Sea Fisheries Since the Second World War," *Polar Record* 13, no. 83 (1966): 163.

111. Albert L. Weeks, *Russia's Life-Saver: Lend-Lease Aid to the U.S.S.R. in World War II* (Lanham, MD: Lexington Books, 2004), 23.

112. Ibid., 152.

113. *Pacific Fisherman*, December 1947, 15.

114. "Made in America—*Alma-Ata*—Russian's Great Floating Cannery," *Pacific Fisherman* (April 1946): 35–36.

CHAPTER TWO

1. John Dower, cited by Jon Halliday, *A Political History of Japanese Capitalism* (New York: Pantheon Books, 1975), xvii.

2. Donald E. Nuechterlein, *Iceland, Reluctant Ally* (Ithaca, NY: Cornell University Press, 1961), 4.

3. Hans W. Weigert, "Iceland, Greenland, and the United States," *Foreign Affairs* 23, no. 1 (October 1944): 113.

4. Nuechterlein, *Iceland, Reluctant Ally*, 24.

5. Ibid., 29.

6. Box 4767, "U.S. Military Requirements in Iceland," May 1, 1946, RG 59, NARA.

7. Einar Benediktsson, "At Crossroads: Iceland's Defense and Security Relations, 1940–2011," Strategic Studies Institute, August 18, 2011, http://www.strategicstudiesinstitute.army.mil/index.cfm/articles/Icelands-Defense-and-Security-Relations-1940–2011/2011/8/18.

8. W. R. Mead, "Renaissance of Iceland," *Economic Geography* 21, no. 2 (1945): 141.

9. Rögnvaldur Hannesson, *Fisheries Mismanagement: The Case of the North Atlantic Cod* (Cambridge, MA: MIT Press, 2004), 63.

10. Hannes Jónsson, *Friends in Conflict: The Anglo-Icelandic Cod Wars and the Law of the Sea* (London: C. Hurst, 1982), 84.

11. H. P. Valtýsson, "The Sea around Icelanders: Catch History and Discards in Icelandic Waters," in *Fisheries Impacts on North Atlantic Ecosystems: Catch, Effort and National/Regional Data Sets,* ed. D. Zeller, R. Watson, and D. Pauly (Vancouver, BC: Fisheries Centre, University of British Columbia, 2001), 52–87.

12. Ibid.

13. Richard Van Cleve and Ralph W. Johnson, *Management of the High Seas Fisheries of the Northeastern Pacific* (Seattle: University of Washington Press, 1963), 63.

14. Jon Th Thor, *British Trawlers in Icelandic Waters: History of British Steam Trawling off Iceland, 1889–1916, and the Anglo-Icelandic Fisheries Dispute, 1896–1897* (Reykjavik: Fjolvi, 1992), 15.

15. Robb Robinson, *Trawling: The Rise and Fall of the British Trawl Fishery* (Exeter: University of Exeter Press, 1996), 105.

16. Nuechterlein, *Iceland, Reluctant Ally,* 4.

17. Robinson, *Trawling,* 108.

18. Jónsson, *Friends in Conflict,* 51.

19. *Commercial Fisheries Review* 8, no. 5 (May 1946): 44.

20. Olafur Björnsson, "Economic Development in Iceland since World War II," *Weltwirtschaftliches Archiv* 98 (1967): 221, http://www.jstor.org/stable/40436434.

21. Nuechterlein, *Iceland, Reluctant Ally,* 11.

22. *Commercial Fisheries Review* 8, no. 5 (May 1946): 44.

23. *Commercial Fisheries Review* (December 1946): 48–49.

24. Box 5114, April 14, 1954, "Fisheries Fund of Iceland," RG59, NARA.

25. *Fish Market News* (June 1945): 46–47.

26. Box 5114, John H. Morgan to Mr. Hickerson, December 3, 1945, RG 59, NARA.

27. Roger W. Gale, *Americanization of Micronesia: A Study of the Consolidation of U.S. Rule in the Pacific* (Washington, D.C.: University Press of America, 1979), 43.

28. Kimie Hara, *Cold War Frontiers in the Asia-Pacific: Divided Territories in the San Francisco System* (New York: Routledge, 2007).

29. Mark R. Peattie, "Nanshin: The 'Southward Advance,' 1931–1941, as a Prelude to the Japanese Occupation of Southeast Asia," in *The Japanese Wartime Empire, 1931–1945,* ed. Ramon H. Myers, Peter Duus, and Mark R. Peattie (Princeton, NJ: Princeton University Press, 1996), 173.

30. Hal M. Friedman, "The Beast in Paradise: The United States Navy in Micronesia, 1943–1947," *Pacific Historical Review* 52, no. 2 (May 1993): 175–76, http://www.jstor.org/stable/3639910.

31. Ronald Rainger, "Science at the Crossroads: The Navy, Bikini Atoll, and American Oceanography in the 1940s," *Historical Studies in the Physical and Biological Sciences* 30, no. 2 (2000): 349–71.

32. Freidman, "The Beast in Paradise," 184.

33. Michael Shaller, *The American Occupation of Japan: The Origins of the Cold War in Asia* (New York: Oxford University Press, 1985), 54.

34. Lt. John Useem, "The American Pattern of Military Government in Micronesia," *American Journal of Sociology* 40, no. 2 (September 1945): 94.

35. Gale, *Americanization of Micronesia,* 42.

36. M. P. Leffler, "National Security and U.S. Foreign Policy," in *Origins of the Cold War: An International History,* ed. M. P. Leffler and D. S. Painter (New York: Rutledge, 1994), 37.

37. Charles J. Weeks Jr., "The New Frontier, the Great Society, and American Imperialism in Oceania," *Historical Review* 71, no. 4 (February 2002): 92.

38. Francis B. Sayre, "American Trusteeship Policy in the Pacific," *Proceedings of the Academy of Political Science* 22, no. 4 (January 1948): 49.

39. Ibid., 51.

40. Gale, *Americanization of Micronesia,* 76.

41. E. C. Weitzell, "Resource Development in the Pacific Mandated Islands," *Journal of Land and Public Utility Economics* 22, no. 3 (August 1946): 199.

42. Ibid., 203.

43. Robert Trumbull, *Paradise in Trust: A Report on Americans in Micronesia, 1946–1958* (New York: William Sloane Associates, 1959), 40.

44. Joseph Waldo Ellison, "The Partition of Samoa: A Study in Imperialism and Diplomacy," *Pacific Historical Review* 8, no. 3 (September 1939): 259–88.

45. Ibid., 260.

46. I. C. Campbell, "Chiefs, Agitators and the Navy: The Mau in American Samoa, 1920–29," *Journal of Pacific History* 44, no. 1 (June 2009): 45.

47. Weitzell, "Resource Development in the Pacific Mandated Islands," 209.

48. George Moskovita, *Living Off the Pacific Ocean Floor* (Corvallis: Oregon State University Press, 2015).

49. "Three fishermen rescued as trawler Treo goes to bottom off Peacock Spit," *Daily Astoria,* December 2, 1940.

50. Moskovita, *Living Off the Pacific Ocean Floor,* 3.

51. Ibid., 4.

52. Moskovita, *Living Off the Pacific Ocean Floor,* 36.

53. Ibid., 37.

54. Sigurd J. Westrheim, "The Soupfin Shark Fishery of Oregon," Research Briefs, Oregon Fish Commission 3, no. 1 (September 1950).

55. Fishery Statistics of Oregon, Oregon Fish Commission, Contribution No. 16 (September 1951): 11.

56. "Fishermen in Astoria look to lucrative catches at sea," *Oregonian,* May 16, 1944.

57. M. E. Stansby, "Opportunities for Small Business in the Fisheries of the Pacific Northwest and Alaska," *Fishery Market News* 6, no. 8 (February 1945): 2–7.

58. Irene Martin and Roger Tetlow, *Flight of the Bumblebee: The Columbia River Packers Association and a Century in the Pursuit of Fish* (Long Beach: Chinook Observer), 197.

59. George Yost Harry Jr., "Analysis and History of the Oregon Otter Trawl Fishery, 1884–1961" (PhD dissertation, University of Washington School of Fisheries, 1956), http://ir.library.oregonstate.edu/xmlui/handle/1957/16899.

60. *Commercial Fisheries Review* 8, no. 10 (October 1946): 4.

61. Dayton Lee Alverson, *Race to the Sea: An Autobiography of a Marine Biologist* (Seattle: iUniverse, 2005).

62. P. A. Moiseev and I. A. Parakostzov, "Some Data about the Ecology of Rockfish of the Northeast Part of the Pacific Ocean," trans. Donald Bevan, *I Voprosy Ikhtiologii (Problems of Ichthyology)* 1, no. 1 (1961).

63. Lawrence Barber, "New trawler passes test," *Oregonian*, May 16, 1945, p. 20.

64. Dayton Alverson and Sigurd J. Westrheim, "A Review of the Taxonomy and Biology of Pacific Ocean Perch and Its Fishery," *Rapports et Process-Verbaux Des Reunions* vol. 150 (1959), 12–27.

65. *Pacific Fisherman*, May 1946, 25–26.

66. "Fishery Resources of the United States," Letter of the Secretary of the Interior, March 1, 1945, 2.

67. Ibid., 4.

68. Ibid., 158.

69. Harry N. Scheiber, "Pacific Ocean Resources, Science, and Law of the Sea: Wilbert M. Chapman and the Pacific Fisheries, 1945–70," *Ecology Law Quarterly* 13, no. 38 (1986): 407.

70. Wilbert M. Chapman, "The Wealth of the Oceans," *Scientific Monthly* 65, no. 3 (1947): 192–97.

71. Box 26, Folder "Dr. Chapman's Report (Pacific Fisheries, etc.)," Chapman to Nick Bez, Dec. 2, 1946, ATA files, SIO.

72. *Newsweek*, May 20, 1946.

73. Scheiber, "Pacific Ocean Resources, Science, and Law of the Sea," 394.

74. William Warner, *Distant Water: The Fate of the North Atlantic Fisherman* (Boston: Little, Brown, 1977), 32.

75. Mary Ann Petrich and Barbara Roje, *The Yugoslav in Washington State: Among the Early Settlers* (Tacoma: Washington State Historical Society, 1984), 59.

76. R. H. Calkins, "Captains of the Pacific Northwest Maritime Industry," *Marine Digest* (September 20, 1952): 25.

77. "The Baron of the Brine," *Time* (November 4, 1945): 92–93.

78. *Pacific Fisherman* (1939): 27.

79. Box 1, folder 8, Bez to John Goodloe, June 16, 1947, Henry Jackson papers, UWSC.

80. http://carmelfinley.wordpress.com/2013/11/27/the-pacific-explorer-and-its-four-fishing-vessels/

81. *Business Week*, August 18, 1945.

82. *Bumble Bee*, August 1947. Columbia River Packers Association, Columbia River Maritime Museum.

CHAPTER THREE

1. Box 4, folder "Tuna meeting, 1949." Presentation to Industry representatives of proposed program of the section of biology and oceanography, 1950, Richard Van Cleve papers, UWSC.

2. *Fish Market News* (June 1945): 46–47.

3. Box 5114, April 14, 1954, "Fisheries Fund of Iceland," RG59, NARA.

4. *Fishery Market News* (June 1945): 46–47.

5. *Commercial Fisheries Review* 10, no. 7 (July 1948): 34.

6. Ibid., 37.

7. *Commercial Fisheries Review,* 10 (12), December 1948, 33.

8. H. P. Valtýsson, "The Sea around Icelanders: Catch History and Discards in Icelandic Waters," in *Fisheries Impacts on North Atlantic Ecosystems: Catch, Effort and National/Regional Data Sets,* ed. D. Zeller, R. Watson, and D. Pauly (Vancouver, BC: Fisheries Centre, University of British Columbia, 2001), 52–87.

9. Box 4, 5114, December 3, 1945, "Iceland bases," RG 59, NARA.

10. Box 4767, May 1, 1946, "U.S. military requirements in Iceland," RG 59, NARA.

11. Box 5114, May 22, 1947, "OMGUS Purchase of Icelandic Fish," RG 59, NARA.

12. David C. Loring, "The United States-Peruvian 'Fisheries' Dispute," *Stanford Law Review* 23, no. 3 (February 1971): 451.

13. Box 5114, April 18, 1947, Reykjavik to Secretary of State, RG 59, NARA.

14. Box 5114, April 28, 1947, Hugh S. Cumming Jr. to Mr. Hickerson, RG 59, NARA.

15. Box 5114, June 12, 1947, "Developments on U.S. purchase of Iceland Fish," RG 59, NARA.

16. Box 5114, May 20, 1947, "OMGUS Purchase of Icelandic Fish," RG 59, NARA.

17. *Commercial Fisheries Review* 10, no. 2 (February 1948): 29.

18. Loring, "The United States-Peruvian 'Fisheries' Dispute," 446.

19. Box 5114, April 18, 1947, Incoming Telegram, RG 59, NARA.

20. Box 5114, May 2, 1947, Outgoing Telegram, RG 59, NARA.

21. Box 5114, August 8, 1947, Incoming Telegram, RG 59, NARA.

22. Box 5114, August 8, 1947, Reykjavik to Secretary of State, RG 59, NARA.

23. Box 5114, September 12, 1947, William C. Trimble to Secretary of State, RG 59, NARA.

24. Box 5114, August 18, 1948, Richard Butrick to John D. Hickerson, RG 59, NARA.

25. Box 5114, June 15, 1949, George H. Butler to Mr. McWilliams, RG 59, NARA.

26. Box 5114, August 18, 1948, Richard P. Butrick to John D. Hickerson, American Legation, RG 59, NARA.

27. *Commercial Fisheries Review* 10, no. 7 (July 1948): 37.

28. Box 18, folder 10, "Report on Iceland," Chapman papers, UWSC.

29. Box 18, folder 10, Chapman papers, UWSC.

30. Box 5114, December 6, 1946, Outgoing telegram, RG 59, NARA.

31. Box 5114, December 10, 1946, Incoming telegram, RG 59, NARA.

32. Alfred E. Eckes Jr., *Opening American's Market: U.S. Foreign Trade Policy since 1776* (Chapel Hill: University of North Carolina Press, 1995), 155.

33. A. A. Berle Jr., "The Marshall Plan in the European Struggle," *Social Research* 15, no. 1 (March 1948): 3.

34. Loring, "The United States-Peruvian 'Fisheries' Dispute," 450.

35. *Commercial Fisheries Review* (January 1949): 33.

36. Reykjavik 178, May 16, 1950, Current Outlook of Iceland's Fishing Industry, NARA.

37. Reykjavik 564, June 27, 1951, Icelandic Summer Herring Fishery.

38. Edwin O. Reischauer, *The United States and Japan* (Cambridge, MA: Harvard University Press, 1950), 241.

39. Robert A. Fearey, *The Occupation of Japan, Second Phase: 1948–50* (New York: McMillan, 1950), 145.

40. F. A. Nicholson, *Note on Fisheries in Japan* (Madras: The Superintendent, Government Press, 1907), 40.

41. *Japan Times and Mail, Japan's Fisheries Industry 1939* (1939), 21.

42. Wilfram Ken Swartz, "Global Maps of the Growth of Japanese Marine Fisheries and Fish Consumption" (master's thesis, University of British Columbia, 2004), 17.

43. Box 8866, July 16, 1947, William Herrington press conference notes, RG 331, NARA.

44. Box 8866, July 16, 1947, Press Conference Notes, RG 331, NARA.

45. *Seattle Post-Intelligencer*, January 27, 1946, Bez files, courtesy of the Bez family.

46. Supreme Commander for the Allied Powers, "Summation of Non-Military Activities in Japan and Korea," no. 28 (January 1948): 117.

47. Ibid.

48. Supreme Commander for the Allied Powers, "Summation of Non-Military Activities in Japan and Korea," no. 42 (November 1948).

49. Box 8875, July 21, 1950, Department of State to the Acting U.S. Political Advisor for Japan, Folder, "Fishing Area Extensions," RG 331, NARA.

50. Box 8866, April 2, 1949, Memorandum, RG 331, NARA.

51. Box 8866, April 30, 1949, Memorandum, RG 331, NARA.

52. Anderson to Herrington, July 22, 1948, Papers of William C. Herrington, University of California Berkeley School of Law, Carton 2D.

53. Box 8866, October 3, 1948, SCAP to Department of Army, RG 331, NARA.

54. *Commercial Fisheries Review* (November 1946): 46–47.

55. William Herrington, "Imported Fish, a Major New England Problem," *Commercial Fisheries Review* 8, no. 2 (February 1946): 1–26.

56. Harry N. Scheiber, "Pacific Ocean Resources, Science, and Law of the Sea: Wilbert M. Chapman and the Pacific Fisheries, 1945–70," *Ecology Law Quarterly* 13, no. 38 (1986): 405–6.

57. Box 26, Folder, "Dr. Chapman's Report (Pacific Explorations)," Chapman to Oregon Senator Guy Cordon, February 14, 1947, ATA files, SIO.

58. Harry N. Scheiber, *Inter-Allied Conflicts and Ocean Law, 1945–53: The Occupation Command's Revival of Japanese Whaling and Marine Fisheries* (Taipei, Taiwan: Institute of European and American Studies, Academia Sinica, 2001).

59. Box 26, Folder "Dr. Chapman's Report (Pacific Fisheries, etc.)," Speech, given to Pacific Division, American Association for the Advancement of Science, at the Univer-

sity of Nevada, Reno, July 18, 1946, "Problems facing the American Tuna Fishery," ATA files, SIO.

60. Chapman to Condon, February 14, 1947, ATA files, SIO.

61. Box 10, Folder 5, December 18, 1946, Memorandum to William Clayton from Monrad Wallgren, Chapman papers, UWSC.

62. Box 27, October 24, 2945, Dean Acheson to Warren Magnuson, Edward Allen Papers, UWSC.

63. Box 50, Folder 21, Memo, Allen papers, UWSC.

64. Box 18, Folder "United Nations fisheries conference," January 30, 1947, Chapman to Harold Coolidge, Allen papers, UWSC.

65. Wilbert McLeod Chapman, "Tuna in the Mandated Islands," *Far Eastern Survey* 15, no. 20 (October 9, 1945): 319.

66. Chapman to Cordon, February 14, 1947, ATA files, SIO.

67. Box 2, Folder "Chapman, W. M., 1940–48," May 2, 1947, Chapman to Montgomery Phister, Van Cleve papers, UWSC.

68. *Pacific Fisherman* (May 1950): 19–21.

69. Box 1, Folder 48, Folder "Chapman, W. M., 1940–48," Jackson papers, UWSC.

70. Harry N. Scheiber, "U.S. Pacific Fishery Studies, 1945–1970: Oceanography, Geopolitics, and Marine Fisheries Expansion," in *The Ocean Sciences: Their History and Relation to Man,* ed. Walter Lenz and Margaret Deacon (Hamburg: Bundesamt für Seeschiffahrt und Hydrographie, 1990), 471–521.

71. Ibid.

72. *Tribune-Sun,* San Diego, July 23, 1947.

73. Box 1, Folder 48, March 8, 1948, Nick Bez to Senator Tobey, Jackson papers, UWSC.

74. *Pacific Fisherman's News* (September 3, 1947): 3.

75. *Pacific Fisherman* (July 1951): 15–22.

76. *Pacific Fisherman* (Yearbook, 1952): 137–43.

77. "*Saipan* and *Tinian* Bid for Tropical Tuna Business," *Pacific Fisherman* (March 1948): 21–22.

78. Box 12, Folder 17, Miller papers, UWSC.

79. *Pacific Fisherman* (December 1947): 15.

80. "No Information," *Pacific Fisherman* (January 1948): 18.

81. Warren Magnuson to George Marshall, April 1, 1947, RG 59, NARA.

82. Ann Hollick, *U.S. Foreign Policy and the Law of the Sea* (Princeton, NJ: Princeton University Press, 1981), 78.

83. Box 26, Folder "Dr. Chapman's Report," June 16, 1948, Chapman to Richard Croker, June 16, 1948, ATA papers, SIO.

84. The International Commission for North Atlantic Fisheries (ICNAF) and the Inter-American Tropical Tuna Commission (IATTC). For more on Chapman's influence on fisheries science, see *All the Fish in the Sea.*

85. Wilbert M. Chapman, "United States Policy on High Seas Fisheries," *Department of State Bulletin* 20, no. 498 (January 16, 1949): 67–80.

86. Box 15, Folder 14, January 3, 1949, Chapman to Judge W. C. Arnold, Chapman papers, UWSC.

87. Wilbert M. Chapman, "United States Policy on High Seas Fisheries," *Department of State Bulletin* 20, no. 498 (January 16, 1949): 67–80.

88. Ibid., 80.

89. *Pacific Fisherman* (January 1949): 17.

90. *Tuna Fisherman Magazine* 2, no. 4 (1948).

91. *Tuna Fisherman Magazine* 1, no. 1 (December 1947).

92. *Seattle Post-Intelligencer*, July 27, 1948.

93. *Portland Oregonian*, December 1951.

94. *Wall Street Journal*, May 4, 1948.

95. *Tuna Fisherman Magazine*, June 1949.

96. *San Bernardino County Sun* (October 7, 1948): 4, http://www.newspapers.com/newspage/49391769/.

97. Chapman to Herrington, February 23, 1949, ATA files, SIO.

98. Box 4, Folder 12, Chapman papers, UWSC.

CHAPTER FOUR

1. Box 153, Folder 39, "The New Age of Exploration," lecture notes, 1949, Revelle papers, SIO.

2. Thomas W. Zeiler, *American Trade and Power in the 1960s* (New York: Columbia University Press, 1992), 22.

3. Box 55, January 30, 1952, Memorandum on Tuna Import Problem, White House Central Files, Truman Presidential Library.

4. Susan Ariel Aaronson, *Trade and the American Dream: A Social History of Postwar Trade Policy* (Lexington: University of Kentucky Press, 1996), 7.

5. Sayuri Shimizu, *Creating People of Plenty: The United States and Japan's Economic Alternatives, 1950–1960* (Kent, OH: Kent State University Press, 2001), 103.

6. Wilbert M. Chapman, "Analysis of the Social and Political Factors in the Development of Fish Production by U.S. Flag Vessels," in *The Future of the Fishing Industry of the United States*, ed. DeWitt Gilbert, vol. 4 (Seattle: University of Washington Publications, March 25–27, 1968), 262–69.

7. Alfred E. Eckes, "Trading American Interests," *Foreign Affairs* 71, no. 4 (Fall 1992): 135.

8. Thomas A. Petit, "The Impact of Imports and Tariffs on the American Tuna Industry," *American Journal of Economics and Sociology* 19, no. 3 (1960): 283.

9. William C. Herrington, "Imported Fish: A Major New England Problem," *Commercial Fisheries Review* 9, no. 2 (February 1946): 3.

10. "U.S. Imports of Groundfish in 1950 Highest Record," *Commercial Fisheries Review* 13, no. 2 (February 1951): 40–41.

11. *Pacific Fisherman* (January 1948): 53.

12. *Pan-American Fisherman* (December 1948): 11–12.

13. "Japanese Canned Tuna Offered in Australia Cheaper than Australian Product," *Commercial Fisheries Review* 17, no. 8 (August 1955): 38.

14. Harold C. Cary, Testimony, U.S. Tariff Commission. Oral Testimony before the U.S. Tariff Commission, December 10, 1948.

15. *San Diego Union Bulletin*, August 22, 1951.

16. Box 55, January 30, 1952, Memo, Carl Corse to Dr. John Steelman, "Tuna Canned in Brine." White House Central Files, Truman Presidential Library.

17. Wilbert M. Chapman, "The Tuna Import Situation," *Pacific Fisherman* 51, no. 10 (October 1951): 27–31.

18. Box 1, Folder 17, December 18, 1951, Chapman to Tatsunosuke Takasaki, Miller Freeman papers, UWSC.

19. *Pacific Fisherman* 32, no. 1 (January 1934): 18.

20. *Pacific Fisherman* (July 1950): 40.

21. *Pacific Fisherman* (October 1951): 27.

22. *Pacific Fisherman* (February 1952): 25.

23. *Pacific Fisherman* (June 1949): 22.

24. Michael Schaller, *The American Occupation of Japan: The Origins of the Cold War in Asia* (New York: Oxford University Press, 1985), 82.

25. Robert A. Fearey, *The Occupation of Japan, Second Phase: 1948–1950* (New York: MacMillan, 1950), 139.

26. Walter LaFeber, *The Clash: U.S.-Japanese Relations through History* (London: W. W. Norton, 1997), 284.

27. Seigen Miyasato, "John Foster Dulles and the Peace Settlement with Japan," in *John Foster Dulles and the Diplomacy of the Cold War*, ed. Richard H. Immerman (Princeton, NJ: Princeton University Press, 1990), 191–212.

28. Michael M. Yoshida, *Japan and the San Francisco Peace Settlement* (New York: Columbia University Press, 1983), 45.

29. Chalmers Johnson, *MITI and the Japanese Miracle, the Growth of Industrial Policy, 1925–1975* (Stanford: Stanford University Press, 1982), 227.

30. Harry N. Scheiber, "Origins of the Abstention Doctrine in Ocean Law: Japanese-U.S. Relations and the Pacific Fisheries, 1937–1958," *Ecology Law Quarterly* 16, no. 1 (1989): 53.

31. Box 12, Folder 27, January 12, 1951, Telegram from Thomas Sandoz to Rep. Wayne Morse, Chapman papers, UWSC.

32. Scheiber, "Origins of the Abstention Doctrine in Ocean Law," 60.

33. Department of State Bulletin, Vol. XXIV, No. 608, 351–52.

34. Ibid.

35. Carton D, March 5, 1951, Chapman to Montgomery Phister, Herrington papers, University of California Berkeley School of Law.

36. *Commercial Fisheries Review* (May 1951): 1–5.

37. Herrington to Chapman, August 30, 1950, Herrington papers, Berkeley.

38. *Commercial Fisheries Review* (May 1951): 5.

39. "The Japanese Fishing Industry, 1928–39 and prospects for 1953," OIR Report no. 4627, Department of State, Division of Research for Far East, Office of Intelligence Research, March 30, 1948.

40. Rachel Carson, "Food from the Sea: Fish and Shellfish of New England," *Conservation Bulletin* 33 (Department of the Interior, Fish and Wildlife Department, 1943), 47.

41. Ibid., 48.

42. Linda Lear, *Rachel Carson: Witness for Nature* (New York: Henry Holt, 1977).

43. Eugene Skolnikoff, *The Elusive Transformation: Science, Technology, and the Evolution of International Politics* (Princeton, NJ: Princeton University Press, 1992), 4.

44. Box 1, Folder 48, "Chapman, W. M., 1940–48," Jackson papers, UWSC.

45. Box 4, Folder "Tuna meeting, 1949," Van Cleve papers, UWSC.

46. http://carmelfinley.wordpress.com/2013/11/27/the-pacific-explorer-and-its-four-fishing-vessels/

47. H. E. Crowther, "Exploratory Fishing," *Fisheries Gazette* 66, no. 13 (n.d.): 106, https://carmelfinley.files.wordpress.com/2014/06/ef-01.jpg.

48. Charles J. Weeks Jr., "The New Frontier, the Great Society, and American Imperialism in Oceania," *Historical Review* 71, no. 4 (February 2002): 97.

49. M. P. Leffler, "National Security and U.S. Foreign Policy," in *Origins of the Cold War: An International History,* ed. M. P. Leffler and D. S. Painter (New York: Rutledge, 1994), 37.

50. Mark D. Merlin and Ricardo M. Gonzalez, "Environmental Impacts of Nuclear Testing in Remote Oceania, 1946–1996," in *Environmental Histories of the Cold War,* ed. J. R. McNeill and Corinna R. Under (Cambridge: Cambridge University Press, 2010), 168.

51. Robert Trumbull, *Paradise in Trust: A Report on Americans in Micronesia, 1946–1958* (New York: William Sloane Associates, 1959).

52. Weeks, "The New Frontier, the Great Society, and American Imperialism in Oceania," 92.

53. Francis B. Sayre, "American Trusteeship Policy in the Pacific," *Proceedings of the Academy of Political Science* 22, no. 4 (January 1948): 49.

54. Ibid., 51.

55. E. C. Weitzell, "Resource Development in the Pacific Mandated Islands," *Journal of Land & Public Utility Economics* 22, no. 3 (August 1946): 199.

56. Congressional Research Service, "The Federal Minimum Wage and American Samoa," May 2007, Order Code RL34013.

57. Peter Wilson, *Aku! The History of Tuna Fishing in Hawaii and the Western Pacific* (Bloomington: Xlibris, 2011), 73.

58. Joanna Poblete-Cross, "Bridging Indigenous and Immigrant Struggles: A Case Study of American Samoa," *American Quarterly* 62, no. 3 (September 2010): 509, doi:10.1353/aq.2010.0017.

59. Michael P. Hamnett and William Sam Pintz, "The Contribution of Tuna Fishing and Transshipment to the Economics of American Samoa, the Commonwealth of the Northern Mariana Islands, and Guam," Pelagic Fisheries Research Program, SOEST 96–05, JIMAR Contribution 96–303.

60. Congressional Research Service, "The Federal Minimum Wage and American Samoa," 3.

61. Box 42, ATA Correspondence, Chapman to John Quimby, Central Labor Council, San Diego, April 10, 1957. "Re Samoa canneries," ATA files, SIO.

62. Tamio Otsu and Ray F. Sumida, "Distribution, Apparent Abundance, and Size Composition of Albacore (*Thunnus alalunga*) Taken in the Longline Fishery Based in American Samoa, 1954–65," *Fisheries Bulletin* 67, no. 1 (July 1968): 48.

63. Robert Gillett, *A Short History of Industrial Fishing in the Pacific Islands* (Food and Agricultural Organization of the United Nations, 2007), http://www.fao.org/docrep/010/ai001e/ai001e00.htm.

64. Otsu and Sumida, "Distribution, Apparent Abundance, and Size Composition of Albacore," 50.

65. *Commercial Fisheries Review* (August 1955): 15.

66. Jean-Marc Fromentin and Joseph E. Powers, "Atlantic Bluefin Tuna: Population Dynamics, Ecology, Fisheries, and Management," *Fish and Fisheries* 6, no. 4 (December 2005): 291.

67. Box 42, ATA Correspondence, Chapman to John Quimby, Central Labor Council, San Diego, April 10, 1957. "Re Samoa canneries," ATA files, SIO.

68. Ibid.

69. Ibid.

70. Ibid.

71. Hannes Jónsson, *Friends in Conflict: The Anglo-Icelandic Cod Wars and the Law of the Sea* (London: C. Hurst, 1982), 56.

72. L. C. Green, "The Anglo-Norwegian Fisheries Case, 1951," *Modern Law Review* 15, no. 3 (1952): 373.

73. Rögnvaldur Hannesson, *Fisheries Mismanagement: The Case of the North Atlantic Cod* (Hoboken: Wiley-Blackwell, 1996), 62.

74. *Commercial Fisheries Review* (September 1952): 34–35.

75. Valur Ingimundarson, "Buttressing the West in the North: The Atlantic Alliance, Economic Warfare, and the Soviet Challenge in Iceland, 1956–1959," *International History Review* 21, no. 1 (March 1999): 87, http://www.jstor.org/stable/40108917.

76. Ibid., 89.

CHAPTER FIVE

1. Christopher M. Weld, "The First Year of Federal Management of Coastal Fisheries: An Outsider's Perspective," *Fisheries* 4, no. 2 (March–April 1979): 13.

2. William Warner, *Distant Water: The Fate of the North Atlantic Fisherman* (Boston: Little, Brown, 1977), 46.

3. *Fishing News*, October 29, 1954.

4. Deitrich Sahrhage and Johannes Lundbeck, *A History of Fishing* (Berlin: Springer, 1992), 125.

5. U.S. Senate Committee on Commerce, "The Postwar Expansion of Russia's Fishing Industry," Report by the Fisheries Research Institute, 88th Congress, 2nd Session, January 1964 (Seattle: University of Washington Press), 6.

6. Antony C. Sutton, *Western Technology and Soviet Economic Development, 1945 to 1965*, vol. 3 (Stanford: Hoover Institution Press, 1973), xxxv.

7. Warner, *Distant Water*, 399.

8. Yu. Yu. Marti, ed., "Soviet Fishery Investigations in the Northwest Atlantic," trans. Israel Program for Scientific Translations (Washington, D.C.: Office of Technical Services, U.S. Dept. of Commerce, 1962), 5.

9. A. T. Pruter, "Soviet Fisheries for Bottomfish and Herring off the Pacific and Bering Sea Coasts of the United States," *Marine Fisheries Review* 38, no. 12 (December 1976): 3.

10. T. S. Sealy, "Soviet Fisheries: A Review," *Marine Fisheries Review* 36, no. 8 (August 1974): 2–22.

11. Ole A. Mathisen and Donald Bevan, *Some International Aspects of Soviet Fisheries* (Athens: Ohio University Press, 1968), 18.

12. Terence Armstrong, "Soviet Sea Fisheries since World War II," *Polar Record* 13, no. 83 (May 1966): 158.

13. Alan Christopher Finlayson, *Fishing for Truth: A Sociological Analysis of Northern Cod Stock Assessments from 1977–1990* (St. John's: Memorial University of Newfoundland, 1944), 32.

14. Sealy, "Soviet Fisheries: A Review," 5.

15. Armstrong, "Soviet Sea Fisheries since World War II," 167.

16. Mathisen and Bevan, *Some International Aspects of Soviet Fisheries*, 46.

17. Armstrong, "Soviet Sea Fisheries since World War II," 170.

18. Mathisen and Bevan, *Some International Aspects of Soviet Fisheries*, 45.

19. Ibid., 48.

20. Vladil Lysenko, *A Crime against the World*, trans. Michael Glenny (London: Victor Gollancz, 1983), 30.

21. Ibid., 38.

22. Ibid., 50.

23. Ibid., 52.

24. Alfred A. Berzin, "Memoirs of a Soviet Whaling Captain," in *Scientific Reports of Soviet Whaling Expeditions in the North Pacific, 1955–1978*, ed. Y. V. Ivashchenko, P. J. Clapham, and R. L. Brownell Jr., trans. Y. V. Ivashchenko (Springfield, VA: U.S. Dept. of Commerce, 2007), NMFS-AFSC-175, http://www.afsc.noaa.gov/techmemos/nmfs-afsc-175.htm.

25. Kurkpatrick Dorsey, *Whales and Nations: Environmental Diplomacy on the High Seas* (Seattle: University of Washington Press, 2013), 153.

26. Berzin, "Memoirs of a Soviet Whaling Captain," 24.

27. Richard E. Ericson, "The Classical Soviet-Type Economy: Nature of the System and Implications for Reform," *Journal of Economic Perspectives* 5, no. 4 (Fall 1991): 11–27.

28. Harvey Levenstein, *Paradox of Plenty: A Social History of Eating in Modern America* (New York: Oxford University Press, 1993), 113.

29. John S. Houtsma, "The Effects of Imports on United States Groundfish Prices" (master's thesis, McGill University, 1971).

30. Paul Josephson, "The Ocean's Hotdog: The Development of the Fish Stick," *Technology and Culture* 49, no. 1 (January 2008): 48.

31. Thomas Zeiler, *American Trade and Power in the 1960s* (New York: Columbia University Press, 1992), 37.

32. Peter A. Larkin, "An Epitaph for the Concept of Maximum Sustained Yield," *Transactions of the American Fisheries Society* 106, no. 1 (1977): 1–11.

33. Margaret E. Dewar, *Industry in Trouble: The Federal Government and the New England Fisheries* (Philadelphia: Temple University Press, 1983), 46.

34. Ibid., 48.

35. Box 45, Folder 12, September 19, 1951, Chapman to Jack Kask, Chapman papers, UWSC.

36. Box 91, Folder "National Labor Management Council on Foreign Trade Policy," August 20, 1947, Chapman to O. R. Strackbein, ATA files, SIO.

37. *Pacific Fisherman* (October 1951): 27–31.

38. *Pacific Fisherman* (November 1951): 21–22.

39. Box 18, December 1954, Statement of Harold F. Cary, Committee, ATA Files, SIO.

40. *Pacific Fisherman* (March 1952): 15–18.

41. *Pacific Fisherman* (March 1952): 17.

42. *Pacific Fisherman* (August 1952): 87.

43. Box 55, January 30, 1952, Memo, Carl Corse to Dr. John Steelman, "Tuna Canned in Brine." White House Central Files, Truman Presidential Library.

44. Box 90, Folder "May 1952," May 15, 1952, Memorandum of Conversation, Proposed Duty on Tuna Imports, Acheson papers. Truman Presidential Library.

45. Wilbert M. Chapman, "The Tuna Import Situation," *Pacific Fisherman* 49, no. 11 (October 1951): 27–31.

46. Box 1, Folder 17, December 18, 1951, Chapman to Tatsunosuke Takasaki, Freeman Papers, UWSC.

47. *Pacific Fisherman* (October 1951): 27.

48. *Pacific Fisherman* (February 1952): 25.

49. Chapman to President of Ecuador, August 15, 1952, Folder, "D. B. Finn, ATA Affair (Osorio) 1952–55." FAO Archives 14 FI 158.

50. *Commercial Fisheries Review* (April 1951): 51.

51. Memorandum of Conversation, Department of State, February 8, 1952, "American tuna fleet fishing problems off coasts of Central America," RG 59.611.206/2–852. NARA.

52. *Commercial Fisheries Review* (February 1953): 63.

53. Box 76, Folder "Peru," November 29, 1951, Smith to Chapman, ATA Files, SIO.

54. Box 76, Folder "Peru," December 12, 1951, Chapman to Smith, ATA Files, SIO.

55. Box 18, December 1954. Statement of Harold F. Cary, Committee on Reciprocity Information, ATA Files, SIO.

56. Harry N. Scheiber, "Pacific Ocean Resources, Science, and Law of the Sea: Wilbert M. Chapman and the Pacific Fisheries, 1945–70," *Ecology Law Quarterly* 13, no. 3 (September 1986): 390.

57. Box 18, Folder "US Tariff Commission," June 26, 1952, The ATA before the U.S. Tariff Commission, ATA files, SIO.

58. *Commercial Fisheries Review* (April 1953): 74–77. *Tuna Fish—Report on Investigation Conducted Pursuant to a Resolution by the Committee on Finance of the United States*, U.S. Department of Commerce.

59. Dewar, *Industry in Trouble*, 48.

60. Box 786, Folder 149-B-2-Fish (1), White House Central Files, Eisenhower Library.

61. Box 94, Folder "Trade Agreements and Tariff Matters, Fish (1)," Subject Series, Eisenhower Presidential Library.

62. Box 786, Folder 149-B-2 Fish (1), June 25, 1953, Eisenhower to Millikin and Reed, White House Central Files, Eisenhower Presidential Library.

63. *Commercial Fisheries Review* (July 1954): 36.

64. Wilbert M. Chapman, "Prepared for Presentation at the Third Annual Conference of the Law of the Sea Institute, June, 1968, University of Rhode Island. On the United States Fish Industry and the 1958 and 1960 United Nations Conference on the Law of the Sea," June 24–27, 1968, University of Rhode Island.

65. Alfred E. Eckes, "Trading American Interests," *Foreign Affairs* 71, no. 4 (Fall 1992): 147.

66. Box 9, Folder "Tuna Fish (2)," Philip Areeda papers, March 17, 1956, White House press release, Eisenhower Presidential Library.

67. OF 624, Folder 122–1 (2), June 4, 1956, White House Press Release, Eisenhower Presidential Library.

68. Box 12, Subject Series, Fish and Wildlife (Fisheries Loans (2)), January 4, 1957, Eisenhower Presidential Library.

69. Dewar, *Industry in Trouble,* 52.

70. J. J. Madruga, ATA President, to James S. Pass, Pass and Seymour, Inc., President, Syracuse, NY, June 28, 1957, ATA files, SIO.

71. Chapman to G. R. Strackbein, National Labor Management, Council on Foreign Trade Policy, December 5, 1958, ATA files, SIO.

CHAPTER SIX

1. E. I. Surkova, "Redfish, Growth and Age," *International Commission for the Northwest Atlantic Fisheries,* special publication no. 3 (1961): 285–90.

2. Ada Epenshade, "A Program for Japanese Fisheries," *Geographical Review* 39, no. 1 (1949): 76.

3. Edwin O. Reischauer, *The United States and Japan* (Cambridge, MA: Harvard University Press, 1950), 247.

4. Box 8875, Folder "Fishing Area Extensions." Far Eastern Commission and Japanese Fishing Areas, November 7, 1950. NARA, RG 331.

5. Deitrich Sahrhage and Johannes Lundbeck, *A History of Fishing* (Berlin: Springer, 1992), 186.

6. Wilbert Chapman, "The Bering Sea Fisheries," *Far Eastern Survey* 22, no. 4 (March 25, 1953): 34.

7. R. K. Jain, *Japan's Postwar Peace Settlements* (Atlantic Highlands, NJ: Humanities Press, 1978), 1–2.

8. Michael Yoshida, *Japan and the San Francisco Peace Settlement* (New York: Columbia University Press, 1983), 100.

9. Thomas W. Zeiler, *American Trade and Power in the 1960s* (New York: Columbia University Press, 1992), 8.

10. Wilfram Ken Swartz, "Global Maps of the Growth of Japanese Marine Fisheries and Fish Consumption" (master's thesis, University of British Columbia, 2004), 19.

11. Hiroshi Kasahara, "Japanese Distant-Water Fisheries: A Review," *Fishery Bulletin* 70, no. 2 (January 1972): 227–82.

12. Swartz, "Global Maps of the Growth of Japanese Marine Fisheries and Fish Consumption," 18.

13. Georg Borgstrom, *Japan's World Success in Fishing* (London: Fishing News Books, 1964), 275.

14. Swartz, "Global Maps of the Growth of Japanese Marine Fisheries and Fish Consumption," 18.

15. Borgstrom, *Japan's World Success in Fishing,* 130.

16. *Tuna Fisherman Magazine* (June 1954): 18.

17. Kurkpatrick Dorsey, *Whales and Nations: Environmental Diplomacy on the High Seas* (Seattle: University of Washington Press, 2013).

18. *Commercial Fisheries Review* (April 1953): 48–49.

19. *Commercial Fisheries Review* (January 1953): 52–56.

20. *Pacific Fisherman* (February 1955): 7.

21. *Pacific Fisherman* (February 1954): 51–52.

22. *Commercial Fisheries Review* (September 1951): 29–30.

23. David L. Howell, *Capitalism from Within: Economy, Society, and the State in a Japanese Fishery* (Berkeley: University of California Press, 1995), 144–45.

24. Swartz, "Global Maps of the Growth of Japanese Marine Fisheries and Fish Consumption," 22.

25. Ole A. Mathisen and Donald Bevan, *Some International Aspects of Soviet Fisheries* (Athens: Ohio University Press, 1968), 35.

26. George Herrfurth, "Walleye Pollock and Its Utilization and Trade," *Marine Fisheries Review* 49, no. 1 (1987): 63.

27. J. N. Tonnessen and A. O. Johnsen, *The History of Modern Whaling*, trans. R. I. Christophersen (London: C. Hurst, 1982), 529.

28. Borgstrom, *Japan's World Success in Fishing*, 236.

29. Swartz, "Global Maps of the Growth of Japanese Marine Fisheries and Fish Consumption," 12.

30. Ibid., 21.

31. T. K. Wilderbuer, G. E. Walters, and R. G. Bakkala, "Yellowfin Sole, *Pleuronectes asper*, of the Eastern Bering Sea: Biological Characteristics, History of Exploitation, and Management," *Marine Fisheries Review* 54, no. 4 (1992): 1–18.

32. Kevin Bailey, *Billion-Dollar Fish: The Untold Story of Alaska Pollock* (Chicago: University of Chicago Press, 2012), 14.

33. Herrfurth, "Walleye Pollock and Its Utilization and Trade," 63.

34. Sunee C. Sonu, "Surimi," NOAA Technical Memorandum, January 1986, NOAA-TM-NMFS-SWR-0 13.

35. Sahrhage and Lundbeck, *History of Fishing*, 189.

36. Swartz, "Global Maps of the Growth of Japanese Marine Fisheries and Fish Consumption," 21.

37. Robert Gillett, *A Short History of Industrial Fishing in the Pacific Islands* (Food and Agricultural Organization of the United Nations, 2007), http://www.fao.org/docrep/010/ai001e/ai001e00.htm.

38. Tamio Otsu and Ray F. Sumida, "Distribution, Apparent Abundance, and Size Composition of Albacore (*Thunnus alalunga*) Taken in the Longline Fishery Based in American Samoa, 1954–65," *Fisheries Bulletin* 67, no. 1 (July 1968): 50.

39. *Pacific Fisherman*, October 1956.

40. Gary D. Sharp, "Tuna Oceanography—An Applied Science," in *Tuna: Physiology, Ecology, and Evolution*, ed. Barbara A. Block and E. Donald Stevens (San Diego: Academic Press, 2001), 361.

41. Sahrhage and Lundbeck, *A History of Fishing*, 189.

42. Yukio Takeuchi, Kazuhiro Oshima, and Ziro Suzaki, "Inference on Nature of

Atlantic Bluefin Tuna off Brazil Caught by the Japanese Longline Fishery around the Early 1960s," *Collective Volume of Scientific Papers of the ICCAT* 63, no. 1 (2009): 186–94.

43. Sayuri Shimizu, *Creating People of Plenty: The United States and Japan's Economic Alternatives, 1959–1960* (Kent, OH: Kent State University Press, 2001).

44. Borgstrom, *Japan's World Success in Fishing*, 276.

45. Swartz, "Global Maps of the Growth of Japanese Marine Fisheries and Fish Consumption," 15.

46. Borgstrom, *Japan's World Success in Fishing*, 207.

47. Swartz, "Global Maps of the Growth of Japanese Marine Fisheries and Fish Consumption," 21.

48. Chapman to G. R. Strackbein, National Labor Management, Council on Foreign Trade Policy, December 5, 1958, ATA files, SIO.

49. David A. Douglas, "The Oregon Shore-Based Cobb Seamount Fishery, 1991–2003: Catch Summaries and Biological Observations," Oregon Department of Fish and Wildlife Information Reports Number 2011–03 (September 2011): 2, https://nrimp.dfw.state.or.us/CRL/Reports/Info/2011-03.pdf.

50. Donald R. Gunderson, *The Rockfish's Warning* (Seattle: University of Washington Press, 2011).

51. Ibid., 45.

52. https://carmelfinley.wordpress.com/2013/05/24/dayton-1 -alverson-my-mentor/

53. Dayton Lee Alverson, *Race for the Sea* (New York: IUniverse), 329.

54. Ibid.

55. https://carmelfinley.wordpress.com/2013/05/24/dayton-1 -alverson-my-mentor/

56. Gunderson, *Rockfish's Warning*.

57. Donald R. Gunderson, "Population Biology of Pacific Ocean Perch (*Sebastes alutus*) Stocks in the Washington–Queen Charlotte Sound Region, and Their Response to Fishing" (PhD dissertation, University of Washington School of Fisheries, 1976), 135.

58. A. T. Pruter, "Soviet Fisheries for Bottomfish and Herring off the Pacific and Bering Sea Coasts of the United States," *Marine Fisheries Review* 38, no. 12 (December 1976): 1–14.

59. Ibid., 4.

60. Owen S. Hamel and Kotaro Ono, "Stock Assessment of Pacific Ocean Perch in Waters off of the U.S. West Coast in 2011," Pacific Fishery Management Council, Portland, September 20, 2011, http://www.pcouncil.org/wp-content/uploads/Pacific_Ocean_Perch_2011_Assessment.pdf.

61. https://carmelfinley.wordpress.com/2012/11/18/flight-over-the-soviet-fleet-1967/.

62. Sahrhage and Lundbeck, *A History of Fishing*, 211.

63. Terence Armstrong, "Soviet Sea Fisheries since World War II," *Polar Record* 13, no. 83 (May 1966): 163.

64. Ibid., 44.

65. Borgstrom, *Japan's World Success in Fishing*, 238.

66. Tonnessen and Johnsen, *History of Modern Whaling*, 584.

67. Robert S. Otto, "History of King Crab Fisheries with Special Reference to the North Pacific Ocean: Development, Maturity, and Senescence," in *King Crabs of the*

World: Biology and Fisheries Management, ed. Bradley G. Stevens (Boca Rotan: CRC Press, 2014), 81–138.

68. Mathisen and Bevan, *Some International Aspects of Soviet Fisheries*, 48.

69. Ibid., 45.

70. Gunderson, *Rockfish's Warning*, 120.

71. Surkova, "Redfish, Growth and Age," 265.

72. "The Postwar Expansion of Russia's Fishing Industry," Committee on Commerce, United States Senate, 88th Congress, 2nd session, January 23, 1964.

73. Edward Wenk Jr., *The Politics of the Ocean* (Seattle: University of Washington Press, 1972), 47.

74. Dean Conrad Allard Jr., *Spencer Fullerton Baird and the U.S. Fish Commission* (New York: Arno Press, 1978), 182.

75. Daniel L. Bottom, "To Till the Water: A History of Ideas in Fisheries Conservation," in *Pacific Salmon and Their Ecosystems: Status and Future Options*, ed. Deanna J Stouder, Peter A. Bisson, and Robert J. Naiman (New York: Chapman & Hall, 1997), 573.

76. Michael Weber, *From Abundance to Scarcity: A History of U.S. Marine Fisheries Policy* (Washington, D.C.: Island Press, 2002), 21.

77. Thomas Wolff, *In Pursuit of Tuna: The Expansion of a Fishing Industry and Its International Ramifications—The End of an Era* (Tempe: Arizona State University Press, 1980), 54.

78. Wenk, *Politics of the Ocean*, 254.

79. Ibid., 21.

80. Ibid., 250.

81. Weber, *From Abundance to Scarcity*, 30.

82. Kevin Bailey, *Billion-Dollar Fish: The Untold Story of Alaska Pollock* (Chicago: University of Chicago Press, 2003), 39.

83. Weber, *From Abundance to Scarcity*, 79.

84. Box 94, Folder "Trade Agreements and Tariff Matters, Fish 94," Memorandum for Dr. Gabriel Hauge, November 23, 1956, Positive Measures to Ease the Groundfish Situation, Subject Series, Eisenhower Presidential Library.

85. Weber, *From Abundance to Scarcity*, 28.

86. Ernst R. Pariser and Christopher J. Corkery, *Fish Protein Concentrate: Panacea for Protein Malnutrition?* (Cambridge, MA: MIT Press, 1978), 19.

87. Ibid., 23.

88. Ibid., xiii.

89. Ibid., 34.

90. National Academy of Sciences, *Oceanography 1966: Achievements and Opportunities* (Washington, D.C., 1967), 2.

CHAPTER SEVEN

1. Warren G. Magnuson, "Opening Address," in *The Future of the Fishing Industry of the United States*, ed. DeWitt Gilbert, vol. 4 (Seattle: University of Washington Publications, March 25–27, 1968).

2. Hannes Jónsson, *Friends in Conflict: The Anglo-Icelandic Cod Wars and the Law of the Sea* (London: C. Hurst, 1982), 56.

3. L. C. Green, "The Anglo-Norwegian Fisheries Case, 1951," *Modern Law Review* 15, no. 3 (1952): 373.

4. Report from Santiago Embassy to State Department, August 20, 1952, RG 59, 398.245-SA/8–2052, NARA.

5. J. N. Tonnessen and A. O. Johnsen, *The History of Modern Whaling*, trans. R. I. Christophersen (London: C. Hurst, 1982), 554.

6. Foreign Service Despatch, Santiago Embassy to State Department, September 5, 1952, RG 59, 398.245 SA/9–552, NARA.

7. State Department Memorandum, UN Conference on the Law of the Sea, August 19, 1957, RG 22, Box 1, E 209, NARA.

8. Ann Hollick, "The Roots of U.S. Fisheries Policy," *Ocean Development and International Law Journal* 5, no. 1 (1978): 84.

9. Preface, *Papers Presented at the International Technical Conference on the Conservation of the Living Resources of the Sea, Rome, 18 April to 10 May 1955* (New York: United Nations Publications, 1956), iii.

10. Box 1538, Folder 398.245, Telegram, Herrington to State Department, May 12, 1955, RG 59 NARA.

11. Ingimundarson, "Buttressing the West in the North: The Atlantic Alliance, Economic Warfare, and the Soviet Challenge in Iceland, 1956–1959," *International History Review* 21, no. 1 (March 1999): 81.

12. Thomas Wolff, *In Pursuit of Tuna: The Expansion of a Fishing Industry and Its International Ramifications—The End of an Era* (Tempe: Arizona State University Press, 1980), 80.

13. Rögnvaldur Hannesson, *The Privatization of the Oceans* (Cambridge, MA: MIT Press, 2004), 34–35.

14. Chapman to John Wedin, Northwest Trawlers Assn., Seattle, July 31, 1958, Re: Geneva Law of the Sea conference, ATA files, SIO.

15. Ingimundarson, "Buttressing the West in the North," 90.

16. Bruce Mitchell, "Politics, Fish, and International Resource Management: The British-Icelandic Cod War," *Geographical Review* 66, no. 2 (April 1976): 133.

17. H. P. Valtýsson, "The Sea around Icelanders: Catch History and Discards in Icelandic Waters," in *Fisheries Impacts on North Atlantic Ecosystems: Catch, Effort and National/Regional Data Sets*, ed. D. Zeller, R. Watson, and D. Pauly (Vancouver, BC: Fisheries Centre, University of British Columbia, 2001), 52–87.

18. Iain Prattis, "Policy Issues of Control, Dependency and Intervention in the North Atlantic Fishery," *Anthropological Quarterly* 53, no. 4 (October 1980): 242–53, http://www.jstor.org/stable/3318107.

19. Morris Davis, *Iceland Extends Its Fisheries Limits: A Political Analysis* (Copenhagen: Scandinavian University Books, 1963), 52.

20. Ingimundarson, "Buttressing the West in the North," 102.

21. Hannesson, *Privatization of the Oceans*, 113.

22. Valtýsson, "The Sea around Icelanders," 71.

23. Olafur Björnsson, "Economic Development in Iceland since World War II," *Weltwirtschaftliches Archiv* 98 (1967): 243.

24. Edward Wenk Jr., *The Politics of the Ocean* (Seattle: University of Washington Press, 1972), 430.

25. Michael Weber, *From Abundance to Scarcity: A History of U.S. Marine Fisheries Policy* (Washington, D.C.: Island Press, 2002), 66.

26. "Report of the Inter-Agency Task Force on New England Fisheries," Papers of Philip Areeda, Box 9, Folder "Fishing Industry-Import Problems," Eisenhower Library.

27. James Acheson, "Coming up Empty: Management Failure of the New England Groundfishery," *Mast* 10, no. 1 (2011): 58.

28. Margaret Dewar, *Industry in Trouble: The Federal Government and the New England Fisheries* (Philadelphia: Temple University Press, 1983), 178.

29. Sara Tjossem, *The Journey to PICES: Scientific Cooperation in the North Pacific* (Juneau: Alaska Sea Grant, 2005).

30. Wenk, *Politics of the Ocean,* 304.

31. "Conclusions: Analysis and Recommendations," *The Future of the Fishing Industry of the United States,* ed. DeWitt Gilbert, vol. 4 (Seattle: University of Washington Publications, March 25–27, 1968), 13–15.

32. Ibid., 16.

33. Wilbert M. Chapman, "Analysis of the Social and Political Factors in the Development of Fish Production by U.S. Flag Vessels," in *The Future of the Fishing Industry of the United States,* ed. DeWitt Gilbert, vol. 4 (Seattle: University of Washington Publications, March 25–27, 1968), 262–69.

34. Ibid., 263.

35. Hiroshi Kasahara, "U.S. Fishing Industry and Related International Activities," in *The Future of the Fishing Industry of the United States,* ed. DeWitt Gilbert, vol. 4 (Seattle: University of Washington Publications, March 25–27, 1968), 241–47.

36. Hiroshi Kasahara, "Management of Fisheries in the North Pacific," *Journal of the Fisheries Research Board of Canada* 30, no. 12, pt. 2 (1973): 2348–60.

37. Jennifer Hubbard, "In the Wake of Politics: The Political and Economic Construction of Fisheries Biology, 1860–1970," *ISIS* 105, no. 2 (June 2014): 364–78, http://www.journals.uchicago.edu/doi/full/10.1086/676572.

38. William L. Black III, "Soviet Fisheries Agreements with Developing Countries: Benefit or Burden?" *Marine Policy* 7, no. 3 (1983): 163–74.

39. Donald Gunderson, *The Rockfish's Warning* (Seattle: University of Washington Press, 2011), 126.

40. David Helvarg, *Blue Frontier: Saving America's Living Seas* (New York: W. H. Freeman, 2001), 177.

41. James P. Walsh, "The Origins and Early Implementation of the Magnuson-Stevens Fishery Conservation and Management Act of 1976," *Coastal Management* 42, no. 5 (2014): 409–25, http://www.tandfonline.com/doi/abs/10.1080/08920753.2014.947227.

42. Weber, *From Abundance to Scarcity,* 86.

43. Dayton Lee Alverson, *The Race to the Sea: An Autobiography of a Marine Biologist* (Seattle: iUniverse, 2005), 441.

44. Weber, *From Abundance to Scarcity,* 89.

45. Alan Longhurst, *The Mismanagement of Marine Fisheries* (Cambridge: Cambridge University Press, 2010).

46. Sidney Holt and L. M. Talbot, "New Principles for the Conservation of Wild Living Resources," *Wildlife Monographs* 59 (1978): 1–33.

47. Christopher M. Weld, "The First Year of Federal Management of Coastal Fisheries: An Outsider's Perspective," *Fisheries* 4, no. 2 (March–April 1979): 13.

48. Adam Langley, Andrew Wright, Glenn Hurry, John Hampton, Transform Aqorua, and Len Rodwell, "Slow Steps Towards Management of the World's Largest Tuna Fishery," *Marine Policy* 33, no. 2 (March 2009): 271.

49. Gary D. Sharp, "Tuna Oceanography—An Applied Science," in *Tuna: Physiology, Ecology, and Evolution,* ed. Barbara A. Block and E. Donald Stevens (San Diego: Academic Press, 2001), 361.

50. "MARCO, the Puretic Power Block, and Purse Seining," http://www.marcoglobal.com/pdf/History_MARCO_FNI.pdf.

51. Rachel A. Schurman, "Tuna Dreams: Resource Nationalization and the Pacific Islanders' Tuna Industry," *Development and Change* 29 (1998): 108.

52. Christopher J. Carr, "Transformation in the Law Governing Highly Migratory Species: 1970 to the Present," in *Bringing New Law to Ocean Waters,* ed. David D. Caron and Harry N. Scheiber (Berkeley: University of California Press, 2004), 61.

53. Kimie Hara, *Cold War Frontiers in the Asia-Pacific: Divided Territories in the San Francisco System* (London: Rutledge, 2007), 189.

54. Michael Pretes and Elizabeth Petersen, "Rethinking Fisheries Policy in the Pacific," *Marine Policy* 28, no. 4 (2004): 297–309.

55. Schurman, "Tuna Dreams," 113.

56. Franziska Torma, "Environment and Development: West-German Fisheries Experts in Thailand," paper presented at a conference of the European Society for Environmental History, Munich, 20–25 August 2013.

57. John G. Butcher, *The Closing of the Frontier: A History of the Marine Fisheries of Southeast Asia, 1850–2000* (Singapore: Institute of Southeast Asian Studies, 2004), 264.

58. Opinion, "Sweatshops Under the American Flag," *New York Times,* May 10, 2002, http://www.nytimes.com/2002/05/10/opinion/sweatshops-under-the-american-flag.html.

59. Pretes and Petersen, "Rethinking Fisheries Policy in the Pacific," 300.

60. Sharp, "Tuna Oceanography—An Applied Science," 355.

61. Dennis M. King and Harry A. Bateman, "The Economic Impact of Recent Changes in the U.S. Tuna Industry" (California Sea Grant Working Paper No. P-T-47, 1985), 24.

62. Irene Martin and Roger Tetlow, *Flight of the Bumblebee: The Columbia River Packers Association and a Century in the Pursuit of Fish* (Long Beach: Chinook Observer, 2012), 197.

63. King and Bateman, "The Economic Impact of Recent Changes in the U.S. Tuna Industry," 21.

64. Quentin Hanich and Martin Tsamenyi, "Managing Fisheries and Corruption in the Pacific Islands Region," *Marine Policy* 33, no. 2 (March 2009): 384.

65. Langley et al., "Slow Steps Towards Management of the World's Largest Tuna Fishery," 274.

66. Hanich and Tsamenyi, "Managing Fisheries and Corruption in the Pacific Islands Region," 387.

67. Langley et al., "Slow Steps Towards Management of the World's Largest Tuna Fishery," 275.

68. Hannah Parris, "Tuna Dreams and Tuna Realities: Defining the Term 'Maximizing Economic Returns from the Tuna Fisheries' in Six Pacific Island Nations," *Marine Policy* 34, no. 1 (January 2010): 107.

CHAPTER EIGHT

1. David Lowenthal, *The Past Is a Foreign Country* (Cambridge: Cambridge University Press, 1985), 39.

2. Richard Van Cleve and Ralph Johnson, *Management of the High Seas Fisheries of the Northeastern Pacific* (Seattle: University of Washington Press, 1963), 5.

3. William L. Black III, "Soviet Fisheries Agreements with Developing Countries: Benefit or Burden?" *Marine Policy* 7, no. 3 (1983): 163–74.

4. Iain Prattis, "Policy Issues of Control, Dependency and Intervention in the North Atlantic Fishery," *Anthropological Quarterly* 53, no. 4 (October 1980): 242–53, http://www.jstor.org/stable/3318107.

5. J. R. McNeill and William H. McNeill, *The Human Web: A Bird's-Eye View of World History* (New York: W.W. Norton, 2003), 310.

6. S. M. Garcia and C. Newton, "Current Situation, Trends, and Prospects in World Capture Fisheries," in *Global Trends, Fisheries Management*, ed. E. K. Pikitch, Daniel D. Huppert, and Michael P. Sissenwine (Bethesda, MD: American Fisheries Society, 1994), 3–27.

7. Alan Longhurst, *The Mismanagement of Marine Fisheries* (Cambridge: Cambridge University Press, 2010).

8. *Pacific Fisherman* (January 25, 1958): 213.

9. Francis T. Christy Jr., "Transitions in the Management and Distribution of International Fisheries," *International Organization* 31, no. 2, Restructuring Ocean Regimes: Implications of the Third United Nations Conference on the Law of the Sea (Spring 1977): 235–65, http://www.jstor.org/stable/2706404.

10. Michael Graham, "Harvests of the Seas," in *Man's Role in Changing the Face of the Earth*, ed. William L. Thomas (Chicago: University of Chicago Press, 1963), 501.

11. Michael Latham, *Modernization as Ideology: American Social Science and Nation Building in the Kennedy Era* (Chapel Hill: University of North Carolina Press, 2000), 4.

12. Samuel P. Hays, *Conservation and the Gospel of Efficiency* (Cambridge, MA: Harvard University Press, 1959), 3.

13. Box 4425, Wilbert M. Chapman, "Report on Activities with Respect to High Seas Fisheries, 1949," RG 59, NARA.

14. Eugene Skolnikoff, *The Elusive Transformation: Science, Technology, and the Evolution of International Politics* (Princeton, NJ: Princeton University Press, 1992), 4.

15. Kristina M. Gjerde and David Freestone, "Unfinished Business: Deep-Sea Fisheries and the Conservation of Marine Biodiversity Beyond National Jurisdictions," *International Journal of Marine and Coastal Law* 19, no. 3 (2004): 213.

16. Gary D. Sharp, "Tuna Oceanography—An Applied Science," in *Tuna: Physiology, Ecology, and Evolution*, ed. Barbara A. Block and E. Donald Stevens (San Diego: Academic Press, 2001), 361.

17. "MARCO, the Puretic Power Block, and Purse Seining," http://www.marcoglobal.com/pdf/History_MARCO_FNI.pdf.

18. Michael Weber, *From Abundance to Scarcity: A History of U.S. Marine Fisheries Policy* (Washington, D.C.: Island Press, 2002), 38.

19. Ibid., 150.

20. C. Bilger, "US-Soviet Fishing Agreement: Treaty Authorizing Soviet Fishing in US Waters," *Marine Policy* 10, no. 1 (1986): 56.

21. Weber, *From Abundance to Scarcity*, 76.

22. Hannes Jónsson, *Friends in Conflict: The Anglo-Icelandic Cod Wars and the Law of the Sea* (London: C. Hurst, 1982), 6.

23. Kevin Bailey, *Billion-Dollar Fish: The Untold Story of Alaska Pollock* (Chicago: University of Chicago Press, 2012), 71.

24. Hilary K. Josephs, "Japanese Investment in the US Fishing Industry and Its Relation to the 200-Mile Law," *Marine Policy* 2, no. 4 (October 1978): 256.

25. Erwin Laurance, "Salmon-Packer Baits World's Fair Hook," *Seattle Times* (March 5, 1962): 13.

26. *Portland Oregonian*, December 1951.

27. *Seattle Times*, January 25, 1955.

28. John Reddin, "Faces of the City: 500 Attend Nick Bez' 'Little Party,'" *Seattle Times*, September 16, 1960.

29. *Pacific Fisherman* (July 1965): 5–6.

30. Josephs, "Japanese Investment in the US Fishing Industry," 264.

31. William L. Black III, "Soviet Fisheries Agreements with Developing Countries: Benefit or Burden?" *Marine Policy* 7, no. 3 (1983): 170.

32. Francis T. Christy Jr., "Transitions in the Management and Distribution of International Fisheries," *International Organization* 31, no. 2, Restructuring Ocean Regimes: Implications of the Third United Nations Conference on the Law of the Sea (Spring 1977): 235–65, http://www.jstor.org/stable/2706404.

33. Vladimir Kaczynski, "The Economics of Eastern Bloc Ocean Policy," *American Economic Review* 69, no. 2, Papers and Proceedings of the Ninety-First Annual Meeting of the American Economic Association (May 1979): 262–65, http://www.jstor/org/stable/1801654.

34. A. T. Pruter, "Soviet Fisheries for Bottomfish and Herring off the Pacific and Bering Sea Coasts of the United States," *Marine Fisheries Review* 38, no. 12 (December 1976): 14.

35. Weber, *From Abundance to Scarcity*, 64.

36. William Warner, *Distant Water: The Fate of the North Atlantic Fisherman* (Boston: Little, Brown, 1977), 58.

37. Ibid., 243.

38. Josephs, "Japanese Investment in the US Fishing Industry," 265.

39. Christy, "Transitions in the Management and Distribution of International Fisheries," 242.

40. Marcus Haward and Anthony Bergin, "The Political Economy of Japanese Distant Water Tuna Fisheries," *Marine Policy* 25, no. 2 (March 2001): 91–101.

41. Sarah Keene Meltzoff and Edward Lipuma, "The Troubled Seas of Spanish Fishermen: Marine Policy and the Economy of Change," *American Ethnologist* 13, no. 4 (November 1986): 681–99.

42. Bruce Mitchell, "Society, Politics, Fish, and International Resource Management: The British Icelandic Cod War," *Geographical Review* 66, no. 2 (April 1976): 127–38, http://www.jstor.org/stable/213576.

43. Vladimir Kaczynski, "Responses and Adjustments of Foreign Fleets to Controls Imposed by Coastal Nations," *Journal of the Fisheries Research Board of Canada* 36, no. 7 (1979): 800–810.

44. G. D. Hurry, M. Hayashi, and J. J. Maguire, "Report of the Independent Review, International Commission for the Conservation of Atlantic Tunas," PLE-106/2008 (September 2008), http://www.iccat.int/Documents/Meetings/Docs/Comm/PLE-106-ENG.pdf.

45. Bailey, *Billion-Dollar Fish*, 79.

46. https://carmelfinley.wordpress.com/2012/09/08/jergen-westrheim-and-rockfish/.

47. George Moskovita, *Living Off the Pacific Ocean Floor* (Corvallis: Oregon State University Press, 2015), 107.

48. Ibid., 124.

49. Donald R. Gunderson, *The Rockfish's Warning* (Seattle: University of Washington Press, 2011), 122.

50. Donald R. Gunderson, "Population Biology of Pacific Ocean Perch (*Sebastes alutus*) Stocks in the Washington–Queen Charlotte Sound Region, and Their Response to Fishing" (PhD dissertation, University of Washington School of Fisheries, 1976).

51. S. J. Westrheim, "Age Determination and Growth of Pacific Ocean Perch (*Sebastes alutus*) in the Northeast Pacific Ocean," *Journal of the Fisheries Research Board of Canada* 30, no. 2 (February 1973): 235–47, doi:10.1139/f73-041.

52. Prattis, "Policy Issues of Control, Dependency and Intervention in the North Atlantic Fishery," 243.

53. Dennis Scarnecchia, "Salmon Management and the Search for Values," *Canadian Journal of Fisheries and Aquatic Sciences* 45, no. 11 (1998): 2044, 2046.

54. Daniel Pauly, *On the Sex of Fish and the Gender of Scientists* (London: Chapman and Hall, 2004), 102.

55. Garrett Hardin, "Tragedy of the Commons," *Science* 162, no. 3859 (December 1, 1968): 1243–48, http://science.sciencemag.org/content/162/3859/1243.full.

56. Jennifer Hubbard, "In the Wake of Politics: The Political and Economic Construction of Fisheries Biology, 1860–1970," *ISIS* 105, no. 2 (June 2014): 365, http://www.journals.uchicago.edu/doi/full/10.1086/676572.

57. E. S. Russell, *The Overfishing Problem* (Cambridge: Cambridge University Press, 1942), 1.

58. Milner B. Schaefer, "The Scientific Basis for a Conservation Program," *Papers Presented at the International Technical Conference on the Conservation of the Living Resources of the Sea: Rome, 18 April to 10 May 1955* (New York: United Nations Publications, 1956), 15.

59. Ibid.

60. Erik M. Poulsen, "Conservation Problems in the Northwestern Atlantic," *Papers Presented at the International Technical Conference on the Conservation of the Living Resources of the Sea: Rome, 18 April to 10 May 1955* (New York: United Nations Publications, 1956), 184.

61. National Science Foundation, *50 Years of Ocean Discovery* (Washington, D.C.: NSF, 2000), 9–21.

62. Naomi Oreskes, "Scaling Up Our Vision," *ISIS* 105, no. 2 (June 2014): 379, 391.

63. Gary S. Morishima and Kenneth A. Henry, "The History and Status of Pacific Northwest Chinook and Coho Salmon Ocean Fisheries and Prospects for Sustainability," in *Sustainable Fisheries Management: Pacific Salmon,* ed. E. Eric Knudsen et al. (Boca Raton: Lewis Publishers, 2000), 219–35.

64. Hubbard, "In the Wake of Politics," 365.

65. Jim Lichatowich, *Salmon Without Rivers: A History of the Pacific Salmon Crisis* (Washington, D.C.: Island Press, 1999).

66. Malin L. Pinsky, Olaf P. Jensen, Daniel Ricard, and Stephen R. Palumbi, "Unexpected Patterns of Fisheries Collapse in the World's Oceans," *Proceedings of the National Academy of Sciences of the United States of America* 108, no. 20 (May 17, 2011): 8317–22.

67. Tim Smith, *Scaling Fisheries: The Science of Measuring the Effects of Fishing, 1855–1955* (Cambridge: Cambridge University Press, 2004).

68. Valur Ingimundarson, "Buttressing the West in the North: The Atlantic Alliance, Economic Warfare, and the Soviet Challenge in Iceland, 1956–1959," *International History Review* 21, no. 1 (March 1999): 80–103, http://www.jstor.org/stable/40108917.

69. Thomas W. Zeiler, *American Trade and Power in the 1960s* (New York: Columbia University Press, 1992), 29.

70. Hannah Parris, "Tuna Dreams and Tuna Realities: Defining the Term 'Maximizing Economic Returns from the Tuna Fisheries' in Six Pacific Island Nations," *Marine Policy* 34, no. 1 (January 2010): 107.

CONCLUSIONS

1. Jon Th Thor, *British Trawlers in Icelandic Waters,* trans. Hilmar Foss (Reykjavik: Fjolvi Publishers, 1992), 235.

2. Elizabeth Mancke, "Early Modern Expansion and the Politicization of Oceanic Space," *American Geographical Society* 89, no. 2 (April 1999): 234.

3. J. C. Perry, *Facing West: Americans and the Opening of the Pacific* (Westport, CT: Praeger, 1994), 153.

4. Immanuel Wallerstein, *The Modern World System: Capitalist Agriculture and the Origins of the European World-Economy in the Sixteenth Century* (New York: Academic Press, 1974).

5. Alan K. Smith, *Creating a World Economy: Merchant Capitalism, Colonialism, and World Trade, 1400–1825* (Boulder: Westview Press, 1991), 6.

6. C. R. Boxer, *The Dutch Seaborne Empire, 1600–1800* (London: Penguin, 1990), 76.

7. Jon Halliday, *A Political History of Japanese Capitalism* (New York: Pantheon, 1975), 54.

8. Ian Urbina, "Sea Slaves: The Human Misery that Feeds Pets and Livestock," *New York Times*, July 27, 2015, http://www.nytimes.com/2015/07/27/world/outlaw-ocean-thailand-fishing-sea-slaves-pets.html.

9. Richard Van Cleve and Ralph W. Johnson, *Management of the High Seas Fisheries of the Northeastern Pacific* (Seattle: University of Washington Press, 1963), 13.

10. Paul R. Josephson, *Industrialized Nature: Brute Force Technology and the Transformation of the Natural World* (Washington, D.C.: Island Press, 2002), 9.

11. Hanna J. Cortner and Margaret Moote, *The Politics of Ecosystem Management* (Washington, D.C.: Island Press, 1999), 14.

12. Jim Lichatowich, *Salmon Without Rivers: A History of the Pacific Salmon Crisis* (Washington, D.C.: Island Press, 1999).

13. S. A. Berkeley et al., "Fisheries Sustainability via Protection of Age Structure and Spatial Distribution of Fish Populations," *Fisheries* 29, no. 8 (July 2004): 23–32, doi:10.1577/1548-8446(2004)29[23:FSVPOA]2.0.CO;2.

14. Daniel Pauly, "One Hundred Million Tonnes of Fish, and Fisheries Research," *Fisheries Research* 25 (1996): 25–38, http://citeseerx.ist.psu.edu/viewdoc/download?doi=10.1.1.179.1957&rep=rep1&type=pdf.

15. Daniel Pauly and Dirk Zeller, "Catch Reconstructions Reveal that Global Marine Fisheries Catches Are Higher than Reported and Declining," *Nature Communications* 7 (January 19, 2016), http://www.nature.com/ncomms/2016/160119/ncomms10244/full/ncomms10244.html, doi:10.1038/ncomms10244.

16. Daniel Pauly, "Diagnosing and Solving the Global Crisis of Fisheries: Obstacles and Rewards," *Cybium* 36, no. 4 (December 2012): 504, http://sfi.mnhn.fr/cybium/numeros/2012/364/02-Pauly823.pdf.

17. Willa Nehlsen, Jack E. Williams, and James A. Lichatowich, "Pacific Salmon at the Crossroads: Stocks at Risk from California, Oregon, Idaho, and Washington," *Fisheries* 16, no. 2 (1991): 4–21, doi: 10.1577/1548-8446(1991)016<0004:PSATCS>2.0.CO;2.

18. Ransom Myers and Boris Worm, "Rapid Worldwide Depletion of Predatory Fish Communities," *Nature* 423 (May 15, 2003): 280–83, doi:10.1038/nature01610.

19. Ellen Pikitch et al., "Ecosystem-Based Fishery Management," *Science* 305 (July 2004): 346–47.

20. Naomi Oreskes, "Scaling Up Our Vision," *ISIS* 105, no. 2 (June 2014): 379–91.

21. Dayton Lee Alverson, *Race to the Sea: An Autobiography of a Marine Biologist* (Seattle: iUniverse, 2005), vii.

22. Berkes, "Shifting Perspectives on Resource Management," 19.

23. Daniel L. Bottom, "To Till the Water: A History of Ideas in Fisheries Conservation," in *Pacific Salmon and Their Ecosystems: Status and Future Options*, ed. Deanna J. Stouder, Peter A. Bisson, and Robert J. Naiman (New York: Chapman and Hill, 1997), 590.

24. Spencer Apollonio and Jacob J. Dykstra, *An Enormous, Immensely Complicated Intervention* (Washington, D.C.: Island Press, 2008), 200.

25. "Capital Construction Fund Program," http://www.nmfs.noaa.gov/mb/financial_services/ccf.htm, accessed May 2016.

26. J. Hentati-Sundberg, K. Fryers-Hellqvist, and A. Duit, "Evidence for Path Dependency in a 100-Year Empirical Study of Swedish Fisheries Policy" (unpublished manuscript).

27. Kurkpatrick Dorsey, *Whales and Nations: Environmental Diplomacy on the High Seas* (Seattle: University of Washington Press, 2013), 287.

28. J. Lubchenco and K. Grorud-Colvert, "Making Waves: The Science and Politics of Ocean Protection," *Science* 350, no. 6259 (October 2015): 382–83, http://science.sciencemag.org/content/350/6259/382.

29. "Goal 14: Life Below Water," http://www.undp.org/content/undp/en/home/sdgoverview/post-2015-development-agenda/goal-14.html.

30. "The 10x20 Initiative," http://www.italyun.esteri.it/rappresentanza_onu/en/comunicazione/cittadini/the-10x20-initiative-rome-march.html.

31. Berkes, "Shifting Perspectives on Resource Management," 14.

32. Owen S. Hamel and Kotaro Ono, "Stock Assessment of Pacific Ocean Perch in Waters off of the U.S. West Coast in 2011," Pacific Fishery Management Council, Portland, September 20, 2011, http://www.pcouncil.org/wp-content/uploads/Pacific_Ocean_Perch_2011_Assessment.pdf.

INDEX